Lecture Notes in Mathematics

Edited by A. Dold and B. Eckmann

1315

Robert A. McCoy
Ibula Ntantu

Topological Properties of Spaces of Continuous Functions

Springer-Verlag
Berlin Heidelberg New York London Paris Tokyo

Authors

Robert A. McCoy
Department of Mathematics
Virginia Polytechnic Institute and State University
Blacksburg, VA 24061, USA

Ibula Ntantu
Department of Mathematics
Middle Tennessee State University
Murfreesboro, TN 37132, USA

Mathematics Subject Classification (1980): 54C35, 54D99, 54E99

ISBN 3-540-19302-2 Springer-Verlag Berlin Heidelberg New York
ISBN 0-387-19302-2 Springer-Verlag New York Berlin Heidelberg

Printing and binding: Druckhaus Beltz, Hemsbach/Bergstr.
2146/3140-543210

TABLE OF CONTENTS

INTRODUCTION

Spaces of functions have been used since the late 19th century to form a framework in which convergence of sequences of functions could be studied. Since then several natural topologies have been frequently used to study function spaces. The purpose of this book is to bring together the techniques used in studying the topological properties of such function spaces and to organize and present the theory in a general setting. In particular, a study is made of C(X,R), the space of all continuous functions from a topological space X into a topological space R.

For almost any natural topology imposed on C(X,R), the topological properties of X and R interact with the topological properties of C(X,R). One of the things which is emphasized is the study of these interactions, especially the deduction of the topological properties of C(X,R) from those of X and R. The two major classes of topologies on C(X,R) which are studied are the set-open topologies and the uniform topologies. Each chapter has a number of exercises, not only about these two classes of topologies, but about other kinds of function space topologies found in the literature. Chapters I, II and III contain basic properties and techniques, as well as classical theory. Chapters IV and V have the characterizations of many topological properties of function spaces. Those in Chapter IV are given in the more general setting of cardinal functions.

The range space throughout this book is denoted by R, and whenever the properties of R are not important for the discussion, C(X,R) is abbreviated as C(X). in order to eliminate pathologies and ensure that C(X,R) is large enough, all spaces are assumed to be completely regular Hausdorff spaces, and R is assumed to contain a nontrivial path. The symbol ω denotes the first infinite ordinal number (which is the set of all natural numbers), and $\mathbb{R}$ is used to indicate the space of real numbers with the usual topology.

Chapter I

FUNCTION SPACE TOPOLOGIES

A concept which plays an important role is that of a network on a space. Let α be a family of subsets of X. A nonempty family β of nonempty subsets of X is an α-network on X provided that for each $A \in \alpha$ and open neighborhood U of A there exists a $B \in \beta$ such that $A \subset B \subset U$. A network on X is an α-network on X where α consists of the singleton subsets of X. A network on X is called a closed (or compact) network on X provided each member is closed (or compact). Similarly a closed (or compact) neighborhood base for X is a neighborhood base for X such that each member is closed (or compact).

1. Set-open Topologies.

If $A \subset X$ and $B \subset R$, then the notation $[A,B]$ is defined by

$$[A,B] = \{f \in C(X,R): f(A) \subset B\}.$$

It is straightforward to check that

$$[A,B_1 \cap B_2] = [A,B_1] \cap [A,B_2], \text{ and}$$

$$[A_1 \cup A_2, B] = [A_1,B] \cap [A_2,B].$$

If $x \in X$ and $B \subset R$, then $[\{x\},B]$ is abreviated as $[x,B]$.

A topology on $C(X,R)$ is called a set-open topology provided there is some closed network α on X such that

$$\{[A,V]: A \in \alpha \text{ and } V \text{ is open in } R\}$$

is a subbase for the topology. In this case the function space having this topology is denoted by $C_\alpha(X,R)$ or $C_\alpha(X)$. In addition, if Y is a subspace of X, then $C_\alpha(Y,R)$ denotes $C_\beta(Y,R)$ where $\beta = \{A \cap Y : A \in \alpha\}$.

For topological spaces X and Y, the notation $X \leq Y$ means that X and Y have the same underlying set and the topology on Y is finer than or equal to the topology on X. With this notation, the following can be established.

Theorem 1.1.1. If α and β are closed networks on X, then $C_\alpha(X) \leq C_\beta(X)$ if and only if every member of α is contained in a finite union of members of β.

Proof. Let p: $[0,1] \to R$ be a path (continuous function) in R such that $p(0) \neq p(1)$, let f_0 be the constant function taking X to p(0), and let $V = R \setminus \{p(1)\}$. Take any $A \in \alpha$. Then $[A,V]$ is a neighborhood of f_0 in $C_\beta(X)$, so that there exists a basic neighborhood $W = [B_1,V_1] \cap \ldots \cap [B_n,V_n]$ of f_0 in $C_\beta(X)$ which is contained in $[A,V]$. Let $B = B_1 \cup \ldots \cup B_n$. To show that $A \subset B$, suppose on the contrary that there exists some $x \in A \setminus B$. Since X is completely regular, there exists a $\phi \in C(X,[0,1])$ such that $\phi(B) = \{0\}$ and $\phi(x) = 1$. Then $p \circ \phi \in W$ while $p \circ \phi \notin [A,V]$, which is a contradiction. This establishes the necessity; the sufficiency is immediate. ■

There are two well-studied examples of set-open topologies. One is the point-open topology, or topology of pointwise convergence, where the closed network on X is the family of all nonempty finite subsets of X. This function space is denoted by $C_p(X,R)$ or $C_p(X)$. On the other hand, the other commonly used set-open topology is the compact-open topology, or topology of compact convergence, where the closed network on X is the family of all nonempty compact subsets of X. This function space is denoted by $C_k(X,R)$ or $C_k(X)$.

The next couple of theorems give facts about the topology of pointwise convergence. The proof of the first fact follows immediately from the definition of the product topology, and the second fact follows from Theorem 1.1.1.

Theorem 1.1.2. The space $C_p(X,R)$ is a dense subspace of R^X with the Tychonoff product topology.

Theorem 1.1.3. If α is any closed network on X, then $C_p(X) \leq C_\alpha(X)$.

Therefore the topology of pointwise convergence is the smallest set-open topology. The largest set-open topology may be obtained by taking the family of all nonempty closed subsets for the closed network. The function space having the largest set-open topology is denoted by $C_w(X,R)$ or $C_w(X)$. These special set-open topologies are then related by

$$C_p(X) \leq C_k(X) \leq C_w(X).$$

These inequalities are only equalities for special X, as given by the following corollary to Theorem 1.1.1.

<u>Theorem 1.1.4.</u> The space $C_p(X) = C_k(X)$ if and only if every compact subset of X is finite; and $C_k(X) = C_w(X)$ if and only if X is compact.

The next theorem establishes the separation properties of set-open topologies.

<u>Theorem 1.1.5.</u> If α is a closed network on X, then $C_\alpha(X)$ is a Hausdorff space. Furthermore, if α is a compact network on X, then $C_\alpha(X)$ is completely regular.

Proof. The first part is immediate, so to show the second part, let $f \in [A,V]$ (a subbasic set suffices since a finite minimum of continuous functions is continuous). Now there exists a $\psi \in C(R,[0,1])$ such that $\psi(f(A)) = \{0\}$ and $\psi(R \setminus V) = \{1\}$. Then define $\phi \in C(C_\alpha(X), [0,1])$ by $\phi(h) = \sup\{\psi(h(a)): a \in A\}$ for each $h \in C_\alpha(X)$. It follows that $\phi(f) = 0$ and $\phi(C_\alpha(X) \setminus [A,V]) = \{1\}$. ∎

Sometimes it is more convenient to work with basic open subsets of the range space rather than with arbitrary open subsets. If some additional assumptions are made about the closed network on X, then it is sufficient to use basic open sets in R to generate the topology on $C_\alpha(X,R)$. A closed network is called <u>hereditarily closed</u> provided that every closed subset of a member is a member.

Theorem 1.1.6. If α is a hereditarily closed, compact network on X and σ is a subbase for R, then $\{[A,S]: A \in \alpha \text{ and } S \in \sigma\}$ is a subbase for $C_\alpha(X,R)$.

Proof. Let $A \in \alpha$, let V be open in R, and let $f \in [A,V]$. For each $a \in A$, there exists a finite subset $\sigma_a \subset \sigma$ such that $f(a) \in \cap\{S: S \in \sigma_a\} \subset V$, and there exists a neighborhood U_a of a in X such that $\overline{U}_a \subset f^{-1}(\cap\{S: S \in \sigma_a\})$. Since A is compact, there exists a finite subset A′ of A such that $S \subset \cup\{U_a: a \in A'\}$. Then define

$$W = \cap\{[A\cap\overline{U}_a,S]: a \in A' \text{ and } S \in \sigma_a\},$$

which clearly contains f. To show that $W \subset [A,V]$, let $g \in W$ and let $x \in A$. Then for some $a \in A'$, $x \in U_a$, so that $g(x) \in \cap\{S: S \in \sigma_a\} \subset V$. ∎

Additional algebraic structures on R induce corresponding structures on C(X,R). For example, if R is a group with operation +, then for each $f,g \in C(X,R)$, $f + g$ is defined by $(f + g)(x) = f(x) + g(x)$ for each $x \in X$. This defines the induced group structure on C(X,R).

Whenever α is a hereditarily closed, compact network on X and R is a locally convex topological vector space, then $C_\alpha(X,R)$ is also a locally convex topological vector space. Part of the proof of this is incorporated in the next theorem.

Theorem 1.1.7. If α is a hereditarily closed, compact network on X and R is a topological group, then $C_\alpha(X,R)$ is a topological group.

Proof. Let the group operation be denoted by + as above, and start with $f - g \in [A,V]$. Then for each $a \in A$, there exist neighborhoods V_a and W_a of f(a) and g(a) such that $V_a - W_a \subset V$. Also for each $a \in A$, there exists a closed neighborhood N_a of a in

X such that $f(N_a) \subset V_a$ and $g(N_a) \subset W_a$. Since A is compact, there exists a finite subset A' of A such that $A \subset \cup\{N_a: a \in A'\}$. Then define

$$S = \cap\{[A \cap N_a, V]: a \in A'\} \text{ and}$$

$$T = \cap\{[A \cap N_a, W_a]: a \in A'\},$$

which are neighborhoods of f and g in $C_\alpha(X,R)$. Now it is easy to check that $S - T \subset [A,V]$. ■

As a result of Theorem 1.1.7, if α is a hereditarily closed, compact network on X and R is a topological group, then $C_\alpha(X,R)$ is homogeneous. In this case it generally suffices to work only with neighborhoods of the zero function f_0. Furthermore, if α is closed under finite unions and if $B = [A_1,V_1] \cap \dots \cap [A_n,V_n]$ is a basic neighborhood of f_0, then $A = A_1 \cup \dots \cup A_n \in \alpha$ and $V = V_1 \cap \dots \cap V_n$ contains 0, so that $f_0 \in [A,V] \subset B$. Therefore in this case it suffices to work with sets of the form [A,V] which contain the zero function. This discussion includes two of the most commonly used function spaces, $C_p(X,R)$ and $C_k(X,R)$.

2. Uniform Topologies. Let α be a closed network on X, and let μ be a compatible (diagonal) uniformity on R. The topology induced on C(X,R) by the uniform structure which is about to be defined on C(X,R) is the same whether a diagonal uniformity is used on R or whether its corresponding covering uniformity is used. So all uniformities are taken as diagonal uniformities.

For each $A \in \alpha$ and $M \in \mu$, define

$$\hat{M}(A) = \{(f,g) \in C(X) \times C(X): \text{ for each } x \in A, (f(x), g(x)) \in M\}.$$

In the case that A = X, set $\hat{M} = \hat{M}(X)$. It is straightforward to check that the family $\{\hat{M}(A): A\in\alpha, M\in\mu\}$ is a subbase for a uniformity on C(X). In fact if α is closed under finite unions, then this family is a base for a uniformity on C(X). The space with the topology induced by the uniformity generated by $\{\hat{M}(A): A\in\alpha, M\in\mu\}$ is denoted by $C_{\alpha,\mu}(X,R)$ or $C_{\alpha,\mu}(X)$. The topology induced in this manner is called the uniform topology on α (with respect to μ) or the topology of uniform convergence on α (with respect to μ). The open sets in $C_{\alpha,\mu}(X)$ can be described as the family of all subsets W of C(X) such that for all $f \in W$, there exist $A_1,\ldots, A_n \in \alpha$ and $M_1,\ldots,M_n \in \mu$ with

$$\hat{M}_1(A_1)[f] \cap \ldots \cap \hat{M}_n(A_n)[f] \subset W,$$

where for each $A \in \alpha$ and $M \in \mu$, $\hat{M}(A)[f]$ is defined by

$$\hat{M}(A)[f] = \{g \in C(X)\colon \ (f,g) \in \hat{M}(A)\}.$$

In the case that $\alpha = \{X\}$, then set $C_\mu(X) = C_{\alpha,\mu}(X)$. The topology on $C_\mu(X)$ is called the uniform topology (with respect to μ) or the topology of uniform convergence (with respect to μ). In this case, $\{\hat{M}\colon M \in \mu\}$ is a base for the uniformity inducing this topology, and a subset W of $C_\mu(X)$ is open provided that for each $f \in W$ there is some $M \in \mu$ such that $\hat{M}[f] \subset W$.

As an illustration of these concepts, the proof is given for the sufficiency of the following theorem. The necessity can be established in a manner similar to the proof of Theorem 1.1.1.

Theorem 1.2.1. If α and β are closed networks on X and μ is a compatible uniformity on R, then $C_{\alpha,\mu}(X) \leq C_{\beta,\mu}(X)$ if and only if every member of α is contained in a finite union of members of β.

Proof. (of sufficiency). Let $A \in \alpha$, $M \in \mu$, and $f \in C(X)$. Then there exist $B_1,\ldots,B_n \in \beta$ such that $A \subset B_1 \cup \ldots \cup B_n$. But $\hat{M}(B_1\cup\ldots\cup B_n) = \hat{M}(B_1) \cap \ldots \cap \hat{M}(B_n)$, so that $\hat{M}(B_1)[f] \cap \ldots \cap \hat{M}(B_n)[f] = (\hat{M}(B_1) \cap \ldots \cap \hat{M}(B_n))[f] = \hat{M}(B_1\cup\ldots\cup B_n)[f] \subset \hat{M}(A)[f]$. ∎

The next fact follows immediately from definition.

<u>Theorem 1.2.2.</u> If α is a closed network on X and μ is a compatible uniformity on R, then $C_{\alpha,\mu}(X) \leq C_{\mu}(X)$.

The relation between set-open topologies and uniform topologies is given by the next fundamental result.

<u>Theorem 1.2.3.</u> If α is a compact network on X and μ is a compatible uniformity on R, then $C_{\alpha}(X) \leq C_{\alpha,\mu}(X)$. If, in addition, α is hereditarily closed, then $C_{\alpha}(X) = C_{\alpha,\mu}(X)$.

Proof. Let $A \in \alpha$, let V be open in R, and let $f \in [A,V]$. For each $a \in A$, there exists an $M_a \in \mu$ such that $M_a[f(a)] \subset V$; choose $N_a \in \mu$ such that $N_a \circ N_a \subset M_a$. Now f(A) is compact, so there exists a finite subset A′ of A such that $f(A) \subset \cup\{N_a[f(a)]: a \in A'\}$. Then define $N = \cap\{N_a: a \in A'\}$. To show that $\hat{N}(A)[f] \subset [A,V]$, let $g \in \hat{N}(A)$ and let $x \in A$. There exists some $a \in A'$ with $f(x) \in N_a[f(a)]$, so that $(f(a),f(x)) \in N_a$. Since $(f(x),g(x)) \in N \subset N_a$, then $(f(a),g(x)) \in N_a \circ N_a \subset M_a$. Therefore $g(x) \in M_a[f(a)] \subset V$, so that $g \in [A,V]$.

For the reverse inequality, let $A \in \alpha$, let $M \in \mu$, and let $f \in C(X)$. Let N be a closed and symmetric element of μ such that $N \circ N \circ N \subset M$. Again since f(A) is compact, there exists a finite subset A′ of A so that $f(A) \subset \cup\{N[f(a)]: a \in A'\}$. For each $a \in A'$, define $A_a = A \cap f^{-1}(N[f(a)])$, which is in α since α is hereditarily closed; also define V_a to be the interior of $(N \circ N)[f(a)]$. Finally define $W = \cap\{[A_a,V_a]: a \in A'\}$, which is open in $C_{\alpha}(X)$. Since V_a contains $N[f(a)]$ for each $a \in A'$, then $f \in W$. To see that $W \subset \hat{M}(A)[f]$, let $g \in W$ and let $x \in A$. There exists some $a \in A'$, with $f(x) \in N[f(a)]$, so that $(f(a), f(x)) \in N$. Also $g(x) \in V_a \subset (N \circ N)[f(a)]$, so that $(f(a),g(x)) \in N \circ N$. Then since N is symmetric, $(f(x),g(x)) \in N \circ N \circ N \subset M$, and thus $g \in \hat{M}(A)[f]$. ∎

In particular, it follows from Theorem 1.2.3 that the compact-open topology is the same as the topology of uniform convergence on compact sets (independent of the uniformity used). Also Theorems 1.2.2 and 1.2.3 give the following.

Theorem 1.2.4. If α is a compact network on X and μ is a compatible uniformity on R, then $C_\alpha(X) \leq C_\mu(X)$.

It is of interest to know when the inequality in Theorem 1.2.4 is an equality. One answer is given by the next theorem.

Theorem 1.2.5. A space X is compact if and only if $C_\mu(X) = C_k(X)$ for every compatible uniformity μ on R.

Proof. If X is compact and $\alpha = \{X\}$, then by Theorem 1.2.1, $C_\mu(X) = C_{\alpha,\mu}(X) \leq C_{k,\mu}(X)$. Also $C_{k,\mu}(X) \leq C_\mu(X)$ by Theorem 1.2.2, so that $C_\mu(X) = C_{k,\mu}(X)$. But Theorem 1.2.3 says that $C_{k,\mu}(X) = C_k(X)$. The converse follows from Theorem 1.2.1. ∎

A special kind of uniform topology is the supremum metric topology. In this case the range space R must have a compatible metric ρ, which can be chosen to be bounded. This metric on R induces a metric $\hat{\rho}$ on C(X) defined by

$$\hat{\rho}(f,g) = \sup\{\rho(f(x),g(x)): x \in X\},$$

which is called the supremum metric. If ρ is complete, then $\hat{\rho}$ is also complete.

The ϵ-balls in R with respect to metric ρ are denoted by $B_\rho(x,\epsilon)$ or $B(x,\epsilon)$, while those in C(X) with respect to metric $\hat{\rho}$ are denoted by the similar notation $B_\rho(f,\epsilon)$ or $B(f,\epsilon)$. Then $\{B_\rho(f,\epsilon)$: $f \in C(X)$ and $\epsilon > 0\}$ is a base for some topology on C(X) called the supremum metric topology. The resulting topological space is denoted by $C_\rho(X,R)$ or $C_\rho(X)$.

Every metric naturally induces a uniformity. It turns out that the supremum metric topology is equal to the uniform topology with respect to the uniformity induced by this metric. The next theorem makes this precise.

<u>Theorem</u> <u>1.2.6</u>. For any space X, if ρ is a compatible bounded metric on R and if μ is the uniformity on R induced by ρ, then $C_\rho(X,R) = C_\mu(X,R)$.

Proof. Let $f \in C(X)$ and $\varepsilon > 0$ be given. For each $\delta > 0$, let $M_\delta = \{(s,t) \in R\times R: \ \rho(s,t) < \delta\}$. Then the family $\{M_\delta: \ \delta>0\}$ is a base for μ. To show that $\hat{M}_{\varepsilon/2}[f] \subset B(f,\varepsilon)$, let $g \in \hat{M}_{\varepsilon/2}[f]$. Then $(f,g) \in \hat{M}_{\varepsilon/2}$, so that for every $x \in X$, $(f(x),g(x)) \in M_{\varepsilon/2}$; or $\rho(f(x),g(x)) < \varepsilon/2$. But then $\hat{\rho}(f,g) \le \varepsilon/2 < \varepsilon$, so that $g \in B(f,\varepsilon)$. This establishes that $C_\rho(X,R) \le C_\mu(X,R)$.

For the reverse inequality, let $f \in C(X)$ and $0<\varepsilon<1$. To show that $B(f,\varepsilon) \subset \hat{M}_\varepsilon[f]$, let $g \in B(f,\varepsilon)$. Then $\hat{\rho}(f,g) < \varepsilon$, so that $\rho(f(x),g(x)) < \varepsilon$ for all $x \in X$. But then $(f(x),g(x)) \in M_\varepsilon$ for all $x \in X$, so that $(f,g) \in \hat{M}_\varepsilon$; and thus $g \in \hat{M}_\varepsilon[f]$. ∎

If α is a closed network on X and ρ is a compatible bounded metric on R, then $C_{\alpha,\rho}(X,R)$ is defined as $C_{\alpha,\mu}(X,R)$, where μ is the uniformity on R induced by ρ. Then for a hereditarily closed, compact network α on X, $C_{\alpha,\rho}(X,R) = C_\alpha(X,R)$. This means that for such α, sets of the following form are basic open sets. For each $A \in \alpha$, $f \in C(X,R)$ and $\varepsilon>0$, define

$$\langle A,f,\varepsilon\rangle = \{g\in C(X,R): \text{ for each } a\in A,\ \rho(f(a),g(a))<\varepsilon\}.$$

When R is metrizable, use of this kind of basic open set in $C_\alpha(X,R)$ is sometimes more convenient.

For a metric space R, the topology on $C_\rho(X,R)$ is dependent on the choice of compatible metric ρ on R. That is, different compatible metrics on R may generate different supremum metric topologies on C(X,R). This is illustrated by the following example.

Example 1.2.7. Let R = $\mathbb{R}$ and let ρ be the usual metric on $\mathbb{R}$ bounded by 1. That is, $\rho(s,t) = \min\{1, |s-t|\}$. Also let σ be the metric on $\mathbb{R}$ defined by

$$\sigma(s,t) = \left|\frac{s}{1+|s|} - \frac{t}{1+|t|}\right|,$$

which is compatible with the usual topology. To prove that $C_\rho(\mathbb{R}) \neq C_\sigma(\mathbb{R})$, let f $\in C(\mathbb{R})$ be the identity function, and for each $n \in \omega$ let $f_n \in C(\mathbb{R})$ be defined by $f_n(x) = x$ if $x < n$ and $f_n(x) = n$ if $x \geq n$. Then for each n, $\hat{\rho}(f,f_n) = 1$; while if $x \geq n$,

$$\sigma(f(x),f_n(x)) = \left|\frac{x}{1+x} - \frac{n}{1+n}\right| = \frac{x-n}{(1+n)(1+x)} < \frac{1}{1+n},$$

so that $\hat{\sigma}(f,f_n) \leq \frac{1}{1+n} < \frac{1}{n}$. Therefore for every n, $B_\sigma(f,1/n)$ is not contained in $B_\rho(f,1)$.

This example also shows that different compatible uniformities on R may generate different uniform topologies on C(X,R). A natural question is: when do compatible uniformities (or metrics) on R generate the same topology on C(X,R)? If X is compact, then by Theorem 1.2.5, all compatible uniformities on R generate the compact-open topology on C(X,R). In particular, if X is compact and ρ is a compatible bounded metric on R, then $C_\rho(X,R) = C_k(X,R)$. On the other hand, if R is compact, then there is only one compatible uniformity on R, so that all compatible uniformities on R (and hence by Theorem 1.2.6, all compatible bounded metrics on R) generate the same topology on C(X,R). Although in this latter case, the topology generated on C(X,R) may not be the compact-open topology.

For a compatible uniformity μ on R, $C_\mu(X,R)$ is homogeneous only in special cases.

In fact, $C_\mu(X,R)$ is homogeneous (a topological group) if and only if X is pseudocompact. However, $C_\mu(X,R)$ is still not a topological vector space unless X is compact.

3. Exercises and Problems for Chapter I.

1. Fine Topology. (Krikorian [1969], Eklund [1978], McCoy [1986b]) Let X be a space, let (R,ρ) be a metric space, and let $\mathbb{R}^+$ be the space of positive real numbers.

(a) For each $f \in C(X,R)$ and $\phi \in C(X,\mathbb{R}^+)$, define

$$B_\rho(f,\phi) = \{g \in C(X,R) : \text{for all } x \in X,\ \rho(f(x),g(x)) < \phi(x)\}.$$

The family $\{B_\rho(f,\phi) : f \in C(X,R) \text{ and } \phi \in C(X,\mathbb{R}^+)\}$ is a base for a topology on $C(X,R)$. This topology is called the fine topology with respect to ρ, and is denoted by $C_{f_\rho}(X,R)$.

(b) For each neighborhood W of the diagonal in $R\times R$, there exists an $\varepsilon \in C(R,\mathbb{R}^+)$ such that for every $t \in R$, $B(t,\varepsilon(t))\times B(t,\varepsilon(t)) \subset W$.

(c) For every compatible uniformity μ on R, $C_\mu(X,R) \leq C_{f_\rho}(X,R)$.

(d) Let μ be a compatible uniformity on R. Then $C_\mu(X,R) = C_{f_\rho}(X,R)$ if and only if X is pseudocompact.

(e) If X is paracompact, then the fine topology is independent of the metric ρ.

(f) If R is a topological group, then $C_{f_\rho}(X,R)$ is a topological group under the natural induced operation. But if R is a topological vector space, then $C_{f_\rho}(X,R)$ is not a topological vector space unless X is compact.

2. Graph Topology. (Naimpally [1966], Poppe [1967] and [1968], Naimpally and Pareek [1970], Hansard [1970]) Let X and R be spaces, and for each $f \in C(X,R)$, let $\Gamma(f)$ be the graph of f (a subset of $X\times R$).

(a) For each open G in $X\times R$, let $F_G = \{f \in C(X,R) : \Gamma(f) \subset G\}$. Then $\{F_G : G \text{ is open in } X\times R\}$ is a base for a topology on $C(X,R)$. This topology is called the graph topology, and is denoted by $C_\gamma(X,R)$.

(b) If μ is any compatible uniformity on R, then $C_\mu(X,R) \le C_\gamma(X,R)$.

(c) If ρ is any compatible bounded metric on R, then $C_{f_\rho}(X,R) \le C_\gamma(X,R)$.

(d) For each X, $C_\gamma(X) = C_k(X)$ if and only if X is compact.

(e) If ρ is a compatible bounded metric on R, then $C_\gamma(X,R) = C_\rho(X,R)$ if and only if X is countably compact.

3. Vietoris Topology. If 2^X denotes the set of all nonempty closed subsets of X, then the Vietoris topology on 2^X has as subbase all sets of one of the following two forms: $\{A \in 2^X$: A and V intersect$\}$ and $\{A \in 2^X$: A is contained in V$\}$, where V is open in X. A function space C(X,R) can be thought of as a subspace of $2^{X \times R}$ by identifying each function with its graph. If X is compact, then this "Vietoris" topology on C(X,R) is the same as the compact-open topology.

Chapter II

NATURAL FUNCTIONS

There are a number of naturally defined functions which operate on function spaces. These natural functions play a useful role in studying the topological properties of function spaces.

1. Injections and Diagonal Functions.

If X and R are spaces, for each $t \in R$ let c_t denote the constant map from X onto t. The injection of R into C(X,R) is the function

$$i: R \to C(X,R)$$

defined by $i(t) = c_t$ for each $t \in R$. It is clear that i is one-to-one. In fact for the appropriate topologies on C(X,R), i is an embedding.

Theorem 2.1.1. Let X and R be any spaces.

(a) If α is any closed network on X, then $i: R \to C_\alpha(X,R)$ is a closed embedding.

(b) If μ is any compatible uniformity on R, then $i: R \to C_\mu(X,R)$ is a closed embedding.

Proof. To show that i is an embedding in part (a), it suffices to show for each $A \in \alpha$ and each open V in R, that $i^{-1}([A,V]) = V$. Now $t \in V$ if and only if $c_t \in [A,V]$, which in turn is true if and only if $t \in i^{-1}([A,V])$.

Likewise, for part (b), it suffices to show for each $t \in R$ and each $M \in \mu$, that $i^{-1}(\hat{M}[i(t)]) = M[t]$. This is true since $s \in M[t]$ if and only if $(t,s) \in M$, which in turn is true if and only if $(i(t),i(s)) \in \hat{M}$, and hence true if and only if $s \in i^{-1}(\hat{M}[i(t)])$.

To establish that i(R) is closed, it suffices to show it is closed in $C_p(X,R)$. Let $f \in C(X,R) \setminus i(R)$, so that there exist $x,y \in X$ with $f(x) \neq f(y)$. Let V and W be disjoint open subsets of R containing f(x) and f(y), respectively. Then $[x,V] \cap [y,W]$ is a neighborhood of f contained in $C(X,R) \setminus i(R)$. ■

Therefore, for any closed hereditary property, it is necessary for R to have the property for $C_\alpha(X,R)$ or $C_\mu(X,R)$ to have the property.

Although in general there is no natural injection from X into C(X,R), there is a natural injection from X into the product of copies of R which is sometimes useful. If F is a subset of C(X,R), define the <u>diagonal function</u>

$$\Delta_F: X \to R^F$$

by $\Delta_F(x)(f) = f(x)$ for every $x \in X$ and $f \in F$. With the product topology on R^F, it is immediate that Δ_F is continuous since the composition of Δ_F with each projection is just an element of F, and is thus continuous.

It is useful to know when the diagonal function is an embedding. For this purpose, a subset F of C(X,R) is said to <u>separate points from closed sets</u> provided that whenever A is closed in X and x is a point of $X \setminus A$, then there is some $f \in F$ such that $f(x) \notin \overline{f(A)}$. If F separates points from closed sets, then it is clear that Δ_F is one-to-one. Actually any dense subset of $C_p(X,R)$ separates points from closed sets.

<u>Theorem 2.1.2.</u> If F is a subset of C(X,R) which separates points from closed sets, then $\Delta_F: X \to R^F$ is an embedding.

Proof. Let U be open in X, and let $x \in U$. Now there exists an $f \in F$ such that $f(x) \notin \overline{f(X \setminus U)}$. Define $V = R \setminus \overline{f(X \setminus U)}$, and let $W = [f,V]$ in R^F (that is, $W = \pi_f^{-1}(V)$). To see that $W \cap \Delta_F(X) \subset \Delta_F(U)$, let $\Delta_F(y) \in W$. Then $g \in f^{-1}(V) = X \setminus f^{-1}((\overline{f(X \setminus U)}) \subset U$. Then $W \cap \Delta_F(X)$ is a neighborhood of $\Delta_F(x)$ contained in $\Delta_F(U)$, making $\Delta_F(U)$ open in $\Delta_F(X)$. ■

<u>2. Composition Functions and Induced Functions.</u> If X, Y and R are spaces, define the <u>composition function</u>

$$\Phi\colon C(X,Y)\times C(Y,R) \to C(X,R)$$

by $\Phi(f,g) = g{\circ}f$ for each $f \in C(X,R)$ and $g \in C(Y,R)$.

<u>Theorem 2.2.1.</u> Let X, Y and R be spaces.

(a) If α is a compact network on X and β is a closed neighborhood base on Y, then $\Phi\colon C_\alpha(X,R)\times C_\beta(Y,R) \to C_\alpha(X,R)$ is continuous.

(b) If X is compact, if μ is a compatible uniformity on Y and if ν is a compatible uniformity on R, then $\Phi\colon C_\mu(X,Y)\times C_\nu(Y,R) \to C_\nu(X,R)$ is continuous.

Proof. For part (a), suppose that $\Phi(f,g) \in [A,W]$, where $A \in \alpha$ and W is open in R. Then $g(f(A)) \subset W$, so that $f(A) \subset g^{-1}(W)$. For every $y \in f(A)$, there exists a neighborhood B_y of y from β which is contained in $g^{-1}(W)$. Since f(A) is compact, there exist $y_1,\dots,y_n \in f(A)$ such that f(A) is contained in the union of the interiors of $B_{y_1},\dots,B_{y_n}$; Let V be this union. Thus if $S = [A,V]\times([B_{y_1},W] \cap \dots \cap [B_{y_n},W])$, then $(f,g) \in S$ and $\Phi(S) \subset [A,W]$.

For part (b), consider $\hat{N}[\Phi(f,g)]$, where $N \in \nu$. First let N' be a symmetric element of ν such that $N'{\circ}N'{\circ}N' \subset N$. For every $x \in X$, $f(x) \in g^{-1}(N'[g(f(x))])$, so that there exists an $M_x \in \mu$ such that $(M_x{\circ}M_x)[f(x)] \subset g^{-1}(N'[g(f(x))])$. Since f(X) is compact, there exist $x_1,\dots,x_n \in X$ such that

$$f(X) \subset M_{x_1}[f(x_1)] \cup \dots \cup M_{x_n}[f(x_n)].$$

Let $M \in \mu$ such that $M{\circ}M \subset M_{x_1} \cap \dots \cap M_{x_n}$. It is now routine to establish that

$\Phi(\hat{M}[f]\times\hat{N}'[g]) \subset \hat{N}[\Phi(f,g)]$. ∎

<u>Corollary 2.2.2.</u> If Y is locally compact, then the composition function $\Phi\colon C_k(X,Y)\times C_k(Y,R) \to C_k(X,R)$ is continuous.

Fixing one of the components of the domain of the composition function results in what is called an <u>induced function</u>. In particular, if $g \in C(Y,R)$, then define the induced function

$$g_*: C(X,Y) \to C(X,R)$$

by $g_*(f) = \Phi(f,g) = g \circ f$ for every $f \in C(X,Y)$. Also if $f \in C(X,Y)$, define the induced function

$$f^*: C(Y,R) \to C(X,R)$$

by $f^*(g) = \Phi(f,g) = g \circ f$ for every $g \in C(Y,R)$.

These induced functions preserve composition in the sense that $(g \circ f)_* = g_* \circ f_*$ and $(g \circ f)^* = f^* \circ g^*$. The proof of the next theorem is straightforward.

Theorem 2.2.3. Let $g \in C(Y,R)$.

(a) Then $g_*: C(X,Y) \to C(X,R)$ is one-to-one if and only if g is one-to-one.

(b) Also if $g_*: C(X,Y) \to C(X,R)$ is onto, then g is onto.

The converse of Theorem 2.2.3.b is in general false, but is true for the special case where g is a retraction from a space Y on to a subspace R of Y.

Theorem 2.2.4. Let $g \in C(Y,R)$.

(a) If α is a closed network on X, then $g_*: C_\alpha(X,Y) \to C_\alpha(X,R)$ is continuous.

(b) If μ is a compatible uniformity on Y, if ν is a compatible uniformity on R and if g is in fact uniformly continuous, then $g_*: C_\mu(X,Y) \to C_\nu(X,R)$ is uniformly continuous.

Proof. For part (a), it suffices to show that $g_*^{-1}([A,W]) = [A,g^{-1}(W)]$ for each $A \in \alpha$ and open W in R. Now $f \in g_*^{-1}([A,W])$ if and only if $g(f(A)) \subset W$, which in turn is true if and only if $f \in [A,g^{-1}(W)]$.

For part (b), let $N \in \nu$. Since g is uniformly continuous, there exists an $M \in \mu$ such that $(g(y_1),g(y_2)) \in N$ whenever $(y_1,y_2) \in M$. Suppose that $f_1, f_2 \in C_\mu(X,Y)$ with $(f_1,f_2) \in \hat{M}$. Then for each $x \in X$, $(f_1(x),f_2(x)) \in M$, so that $(g(f_1(x)),g(f_2(x))) \in$

N. Therefore $(g_*(f_1),g_*(f_2)) \in \hat{N}$, showing that g_* is uniformly continuous. ∎

The uniform continuity in Theorem 2.2.4.b cannot be replaced by continuity alone. For example, for R = $\mathbb{R}$, if μ is the uniformity induced by the usual metric on $\mathbb{R}$ and if $g \in C(\mathbb{R})$ is defined by $g(x) = x^2$, then $g_*\colon C_\mu(\mathbb{R}) \to C_\mu(\mathbb{R})$ is not continuous.

<u>Theorem</u> <u>2.2.5.</u> Let $g \in C(Y,Z)$ be an embedding.

(a) If α is a closed network on X, then $g_*\colon$ $C_\alpha(X,Y) \to C_\alpha(X,R)$ is an embedding.

(b) If μ is a compatible uniformity on Y, if ν is a compatible uniformity on R and if g is a uniform embedding (both g and g^{-1} are uniformly continuous), then $g_*\colon$ $C_\mu(X,Y) \to C_\nu(X,R)$ is a uniform embedding.

Proof. The proof of each part is similar to that of the corresponding part in Theorem 2.2.4. The proof of part (a) is given for illustration. It suffices to show that g_* is open onto its image. Since g_* is one-to-one, consider a subbasic open set $[A,V]$ of $C_\alpha(X,Y)$. Since g is an embedding, there is an open set W in R with $W \cap g(X) = g(V)$. By the proof of Theorem 2.2.4.a, $g_*^{-1}([A,W]) = [A,g^{-1}(W)] = [A,V]$. But then $g_*([A,V]) = [A,W] \cap g_*(C_\alpha(X,Y))$, which is open in the image of g_*. ∎

The other kind of induced function is perhaps more useful. It possesses an interesting duality which can be applied to establishing topological properties of function spaces. To this end, define a function to be <u>almost</u> <u>onto</u> if its image is a dense subset of its range. The analog of Theorem 2.2.3 is the following.

<u>Theorem</u> <u>2.2.6</u>. Let $f \in C(X,Y)$.

(a) Then $f^*\colon$ $C(Y) \to C(X)$ is one-to-one if and only if f is almost onto.

(b) If α is a closed network on X and $f^*\colon$ $C(Y) \to C_\alpha(X)$ is almost onto, then f is one-to-one.

(c) If α is a compact network on X and f is one-to-one, then f^*: $C(Y,\mathbb{R}) \to C_\alpha(X,\mathbb{R})$ is almost onto.

Proof. For the sufficiency of part (a), suppose $g_1,g_2 \in C(Y)$ with $f^*(g_1) = f^*(g_2)$, and let $y \in f(X)$. Then for some $x \in X$, $y = f(x)$ and $g_1(y) = g_1(f(x)) = f^*(g_1)(x) = f^*(g_2)(x) = g_2(f(x)) = g_2(y)$. Since f(X) is dense in Y, then $g_1 = g_2$.

For the necessity of part (a), suppose that there exists a $y \in Y \setminus \overline{f(X)}$. Let $p \in C([0,1])$ be a path in R so that $p(0) \neq p(1)$. Now the continuous function mapping $\overline{f(X)}$ onto $\{0\}$ and y to 1 has an extension $\phi \in C(Y,[0,1])$. If $g = p \circ \phi$ and c is the constant map taking Y onto $\{p(0)\}$, then for each $x \in X$, $g(f(x)) = p(0) = c(f(x))$. But then $f^*(g) = f^*(c)$, so that f^* is not one-to-one.

For part (b), let x_1 and x_2 be distinct elements of X. Then there is some $h \in C(X)$ with $h(x_1) \neq h(x_2)$, and there exist disjoint neighborhoods V and W of $h(x_1)$ and $h(x_2)$ in R. Let $S = [x_1,V] \cap [x_2,W]$, which is a neighborhood of h in $C_\alpha(X)$. Then since f^* is almost onto, there is some $g \in C(Y)$ with $f^*(g) \in S$. This means that $g(f(x_1)) \in V$ and $g(f(x_2)) \in W$, so that $f(x_1) \neq f(x_2)$.

Finally for part (c), let $S = [A_1,W_1] \cap \dots \cap [A_n,W_n]$ be a basic open set in $C_\alpha(X,\mathbb{R})$ containing some element h. Define $A = A_1 \cup \dots \cup A_n$, and let $B = f(A)$. Now $f|_A$ is a homeomorphism from A onto B. Then $h \circ (f|_A)^{-1}$: $B \to \mathbb{R}$ has an extension $g \in C(Y,\mathbb{R})$. Since for each $x \in A$, $f^*(g)(x) = g(f(x)) = h(x)$, then $f^*(g) \in S$. ■

In general, f^*: $C(Y) \to C(X)$ is not onto, but a special case in which it is onto is when X is a C-embedded subspace of Y, f is the inclusion map, and $R = \mathbb{R}$.

In order to help formulate the continuity properties of this kind of induced function, some terminology is useful. If α is a closed network on X and if f: $X \to Y$ is a function, define $f(\alpha) = \{f(A) : A \in \alpha\}$. Then α <u>can be approximated by</u> β provided that β

is an α-network.

Theorem 2.2.7. Let $f \in C(X,Y)$, let α be a closed network on X, and let β be a closed network on Y.

(a) If $f(\alpha)$ can be approximated by β, then $f^*: C_\beta(Y) \to C_\alpha(X)$ is continuous.

(b) If f is onto and β can be approximated by $f(\alpha)$, then $f^*: C_\beta(Y) \to C_\alpha(X)$ is open onto its image.

(c) If f is onto and each of $f(\alpha)$ and β can be approximated by the other, then $f^*: C_\beta(Y) \to C_\alpha(X)$ is an embedding.

Proof. For part (a), suppose that $f^*(g) \in [A,V]$, where $g \in C_\beta(Y)$, $A \in \alpha$, and V is open in R. There exist $B_1,\ldots,B_n \in \beta$ such that $f(A) \subset B_1 \cup \ldots \cup B_n \subset g^{-1}(V)$. Then $S = [B_1,V] \cap \ldots \cap [B_n,V]$ is a neighborhood of g in $C_\beta(Y)$ such that $f^*(S) \subset [A,V]$.

For part (b), since f^* is one-to-one by Theorem 2.2.6.a, it suffices to use a subbasic open set in $C_\beta(Y)$. So let $g \in [B,V]$, where $B \in \beta$ and V is open in R. Then there exist $A_1,\ldots,A_n \in \alpha$ such that $B \subset f(A_1) \cup \ldots \cup f(A_n) \subset g^{-1}(V)$. Define $T = [A_1,V] \cap \ldots \cap [A_n,V]$, which is a neighborhood of $f^*(g)$ in $C_\alpha(X)$. To see that $T \cap f^*(C_\beta(Y)) \subset f^*([B,V])$, let $f^*(g') \in T$ for some $g' \in C_\beta(Y)$. Then $g'(B) \subset g'(f(A_1) \cup \ldots \cup f(A_n)) = g'(f(A_1)) \cup \ldots \cup g'(f(A_n)) \subset V$.

Part (c) follows from parts (a) and (b) and from Theorem 2.2.6.a. ∎

In order to rephrase Theorem 2.2.7 in terms of familiar topologies, define $f \in C(X,Y)$ to be a k-covering (compact-covering) map provided that each compact subset of Y is contained in (equal to) the image of some compact subset of X.

Corollary 2.2.8. Let $f \in C(X,Y)$.

(a) Then $f^*: C_p(Y) \to C_p(X)$ is continuous, and is an embedding if f is onto.

(b) Also $f^*: C_k(Y) \to C_k(X)$ is continuous, and is an embedding if f is a k-covering map.

One can show that if f^*: $C_p(Y,\mathbb{R}) \to C_p(X,\mathbb{R})$ is an embedding then f must be onto, and if f^*: $C_k(Y,\mathbb{R}) \to C_k(X,\mathbb{R})$ is an embedding then f must be a k-covering map. In fact the converse of each part of Theorem 2.2.7 is true for $R = \mathbb{R}$ and compact α and β (cf. Exercise 2).

Theorem 2.2.9. Let $f \in C((X,Y)$, and let μ be a compatible uniformity on R. Then f^*: $C_\mu(Y) \to C_\mu(X)$ is continuous. Furthermore, if f is almost onto, then f^* is an embedding.

Proof. Let $g \in C_\mu(Y)$, and let $M \in \mu$. It is straightforward to check that $f^*(\hat{M}[g]) \subset \hat{M}[f^*(g)]$, which establishes the continuity of f^*. To obtain that f^* is open onto its image, it suffices to find an $N \in \mu$ such that $\hat{N}[f^*(g)] \cap f^*(C_\mu(Y)) \subset f^*(\hat{M}[g])$. Such an N can be found by taking a symmetric element of μ with $N \circ N \circ N \subset M$. To check that this works, let $h \in C_\mu(Y)$ such that $f^*(h) \in \hat{N}[f^*(g)]$, and let $y \in Y$. There exist neighborhoods V and W of y in Y such that $V \subset g^{-1}(N[g(y)])$ and $W \subset h^{-1}(N[h(y)])$. Since f is almost onto, there exists an $x \in X$ such that $f(x) \in V \cap W$. Then $(f^*(g)(x),f^*(h)(x)) \in N$, so that $(g(f(x)), h(f(x))) \in N$. Also $(g(y),g(f(x))) \in N$ and $(h(y),h(f(x))) \in N$. Then $(g(y),h(y)) \in N \circ N \circ N \subset M$, so that $h \in \hat{M}[g]$. ∎

Finally, the following condition on f ensures that any of the above embeddings is a closed embedding.

Theorem 2.2.10. If $f \in C(X,Y)$ is a quotient map, then $f^*(C(Y))$ is a closed subset of $C_p(X)$.

Proof. Let $g \in C(X) \setminus f^*(C(Y))$. By way of contradiction, suppose that whenever $x,z \in X$ with $g(x) \neq g(z)$, then $f(x) \neq f(z)$. Define h: $Y \to R$ as follows. If $y \in Y$, then $g(f^{-1}(y))$ is a singleton set; let $h(y)$ be the element in this set. Let W be open in R. Then $f^{-1}(h^{-1}(W)) = g^{-1}(W)$, so that $h^{-1}(W)$ is open in Y, and hence h is continuous. Now $g = h \circ f$, so that $g \in f^*(C(Y))$, which is a contradiction. Therefore there exist $x,z \in X$ such that $g(x) \neq g(z)$ while $f(x) = f(z)$. Let U and V be disjoint neighborhoods of $g(x)$ and $g(z)$, respectively. Then $g \in [x,U] \cap [z,V]$, which is open in $C_p(X)$. Finally, if $h \in [x,U] \cap [z,V]$, then $h(x) \neq h(z)$, and thus $h \notin f^*(C(Y))$. ∎

The topology of pointwise convergence used in Theorem 2.2.10 may of course be replaced by any larger topology.

3. __Evaluation Functions.__ If X and R are spaces, the __evaluation function__

$$e\colon X \times C(X,R) \to R$$

is defined by $e(x,f)=f(x)$ for each $x \in X$ and $f \in C(X)$.

The evaluation function may be expressed in terms of a composition function. Let 1 denote the topological space consisting of a single element. Also let $i_X\colon X \to C(1,X)$ and $j_R\colon R \to C(1,R)$ be the injections. If id denotes the identity map, then

$$i_X \times id\colon X \times C(X,R) \to C(1,X) \times C(X,R)$$

is defined by $(i_X \times id)(x,f) = (i_X(x),f)$ for every $x \in X$ and $f \in C(X,R)$. Finally, let

$$i_X \times id\colon X \times C(X,R) \to C(1,X) \times C(X,R)$$

is defined by $(i_X \times id)(x,f) = (i_X(x),f)$ for every $x \in X$ and $f \in C(X,R)$. Finally, let

$$\Phi\colon C(1,X) \times C(Y,R) \to C(1,R)$$

be the composition function.

__Theorem 2.3.1.__ The evaluation function can be written as $e = i_R^{-1} \circ \Phi \circ (i_X \times id)$.

Proof. Let $x \in X$ and $f \in C(X,R)$. Then $i_R^{-1} \circ \Phi \circ (i_X \times id)(x,f) = i_R^{-1} \circ \Phi(c_x,f) = i_R^{-1} \circ f \circ c_x = i_R^{-1} \circ c_{f(x)} = i_R^{-1} \circ i_R(f(x)) = f(x) = e(x,f)$. ∎

A sufficient condition for the evaluation function to be continuous now follows from

Theorems 2.1.1, 2.2.1 and 2.3.1.

Theorem 2.3.2. Let X and R be spaces.

(a) If α is a closed neighborhood base on X, then e: $X \times C_\alpha(X,R) \to R$ is continuous.

(b) If μ is a compatible uniformity on R, then e: $X \times C_\mu(X,R) \to R$ is continuous.

Corollary 2.3.3. If X is locally compact, then e: $X \times C_k(X,R) \to R$ is continuous.

If X and R are spaces and if $x \in X$, then define the evaluation function at x

$$e_x: C(X,R) \to R$$

by $e_x(f) = e(x,f) = f(x)$ for every $f \in C(X,R)$.

Since for each $x \in X$ and open V in R, $e^{-1}(V) = [x,V]$, then the continuity of e_x: $C_p(X,R) \to R$ follows. Therefore e_x will be continuous for all of the topologies studied on C(X,R).

The map e_x: $C(X,R) \to R$ has an inverse relationship to the injection i: $R \to C(X,R)$ given by the following theorem.

Theorem 2.3.4. Let i: $R \to C(X,R)$ be the injection and let e_x: $C(X,R) \to R$ be the evaluation function at $x \in X$. Then $e_x \circ i$ is the identity on R. Furthermore, if α is any closed network on X, then $i \circ e_x$ is a retraction from $C_\alpha(X,R)$ onto i(R).

Proof. For the first part, if $y \in R$, then $e_x \circ i(y) = e_x(c_y) = c_y(x) = y$. For the second part, if $f \in i(y)$ then $i \circ e_x(f) = i(f(x)) = c_{f(x)} = f$. ∎

Therefore if α is a closed network on X or μ is a compatible uniformity on R, then R may be thought of as a retract of $C_\alpha(X,R)$ or of $C_\mu(X,R)$. In particular, for any property which is preserved by continuous functions, it is necessary that R have this property in order that $C_\alpha(X,R)$ or $C_\mu(X,R)$ have this property.

Even though X cannot be naturally embedded in $C_\alpha(X,R)$, the evaluation functions can be used to describe a way in which X can be naturally embedded in $C_\beta(C_\alpha(X,R),R)$ for certain α and β.

First, restate the definition of the diagonal function $\Delta: X \to R^{C(X,R)}$ as follows. For each $x \in X$, $\Delta(x) = e_x$, where e_x is the evaluation function at x.

If α is a closed network on X or μ is a compatible uniformity on R, then because of Theorem 1.1.2, Δ can be considered as a continuous function from X into $C(C_\alpha(X,R),R)$ or $C(C_\mu(X,R),R)$. From Theorem 2.1.2 it follows that $\Delta: X \to C_p(C_\alpha(X,R),R)$ and $C_p(C_\mu(X,R),R)$ are embeddings. In fact the topology of pointwise convergence can sometimes be strengthened here, as given by the next theorem.

Theorem 2.3.5. Let X and R be spaces.

(a) If α is a closed neighborhood base for X and β is a compact network on $C_\alpha(X,R)$, then $\Delta: X \to C_\beta(C_\alpha(X,R),R)$ is an embedding.

(b) If μ is a compatible uniformity on R and β is a compact network on $C_\mu(X,R)$, then $\Delta: X \to C_\beta(C_\mu(X,R),R)$ is an embedding.

Proof. For each part it suffices to show that Δ is continuous. So let $x \in X$, let $B \in \beta$ and let V be open in R such that $\Delta(x) \in [B,V]$. Then for every $f \in B$, $f(x) \in V$.

For part (a), for every $f \in B$, there exists an $A_f \in \alpha$ such that A_f is a neighborhood of x contained in $f^{-1}(V)$. Since B is compact, there exist $f_1,\dots,f_n \in B$ with

$$B \subset [A_{f_1},V] \cup \dots \cup [A_{f_n},V].$$

Define $A = A_{f_1} \cap \dots \cap A_{f_n}$, which is a neighborhood of x. Also $\Delta(A) \subset [B,V]$, which establishes the continuity of Δ at x.

For part (b), for every $f \in B$, there exists an $M_f \in \mu$ such that $M_f[f(x)] \subset V$. Let

$N_f \in \mu$ be such that $N_f \circ N_f \subset M_f$, and let U_f be a neighborhood of x in X such that $f(U_f) \subset N_f[f(x)]$. Since B is compact, there exist $f_1, \dots, f_n \in B$ with

$$B \subset \hat{N}_{f_1}[f_1] \cup \dots \cup \hat{N}_{f_n}[f_n].$$

Define $U = U_{f_1} \cap \dots \cap U_{f_n}$, which is a neighborhood of x. It is straightforward to check that $\Delta(U) \subset [B,V]$ as desired. ∎

It follows from this theorem that if X is locally compact then $\Delta: X \to C_k(C_k(X))$ is an embedding. This embedding is in fact a closed embedding (cf. Exercise 3).

As an application of the induced functions from section 2, the next theorem extends this previous result to spaces X which are k-spaces. A space X is a <u>k-space</u> provided that if A is a subset of X such that the intersection of A with each compact subset of X is closed, then A must be closed. An equivalent definition is that X is k-space if and only if it is a quotient space of some locally compact space (the disjoint topological sum of all the compact subspaces of X).

<u>Theorem 2.3.6.</u> If X is a k-space, then $\Delta: X \to C_k(C_k(X))$ is an embedding.

Proof. Let $q: Z \to X$ be a quotient map, where Z is locally compact. To show that Δ is continuous, it suffices to show that $\Delta \circ q$ is continuous. Let $\Delta' : Z \to C_k(C_k(Z))$ be the diagonal function on Z, which is continuous by the comment after Theorem 2.3.5. Let $q^* : C_k(X) \to C_k(Z)$ and $q^{**} : C_k(C_k(Z)) \to C_k(C_k(X))$ be the induced and second induced maps, which are continuous by Corollary 2.2.8.b. It remains only to show that $\Delta \circ q = q^{**} \circ \Delta'$. If $z \in Z$ and $f \in C_k(X)$, then $q^{**} \circ \Delta'(z)(f) = \Delta'(z) \circ q^*(f) = \Delta'(z)(f \circ q) = f(q(z)) = \Delta \circ q(z)(f)$. ∎

<u>4.</u> <u>Product Functions and Sum Functions</u>. These final two sections of the chapter deal with the exponential properties of function spaces.

Let $\mathcal{R}$ be a family of spaces, and let $\Pi\mathcal{R}$ denote the cartesian product of the spaces in $\mathcal{R}$ with the product topology. For each $R \in \mathcal{R}$, let $\pi_R: \Pi\mathcal{R} \to R$ be the natural projection. Also if X is a space, then let $\mathcal{R}^X$ denote the family $\{R^X: R \in \mathcal{R}\}$.

Define the <u>product function</u>

$$P: (\Pi\mathcal{R})^X \to \Pi(\mathcal{R}^X)$$

by $\pi_{R^X}(P(f)) = \pi_R \circ f$ for each $f \in (\Pi\mathcal{R})^X$ and $R \in \mathcal{R}$.

<u>Theorem 2.4.1.</u> The product function $P: (\Pi\mathcal{R})^X \to \Pi(\mathcal{R}^X)$ is a bijection.

Proof. The goal is to show that P has as inverse, the function

$$P': \Pi(\mathcal{R}^X) \to (\Pi\mathcal{R})^X$$

defined by $\pi_R \circ P'(g) = \pi_{R^X}(g)$ for each $g \in \Pi(\mathcal{R}^X)$ and $R \in \mathcal{R}$. First let $f \in (\Pi\mathcal{R})^X$. For each $R \in \mathcal{R}$, $\pi_R \circ P' \circ P(f) = \pi_{R^X}(P(f)) = \pi \circ f$, so that $P' \circ P(f) = f$. A similar argument shows that $P \circ P'(g) = g$ for $g \in \Pi(\mathcal{R}^X)$. ∎

If X is a space and $\mathcal{R}$ is a family of spaces, then let $C(X,\mathcal{R})$ denote the family $\{C(X,R) : R \in \mathcal{R}\}$. The next theorem follows from the fact that a function $f \in (\Pi\mathcal{R})^X$ is continuous if and only if $\pi_R \circ f$ is continuous for each $R \in \mathcal{R}$.

<u>Theorem 2.4.2.</u> If $P: (\Pi\mathcal{R})^X \to \Pi(\mathcal{R}^X)$ is the product function, then $P(C(X,\Pi\mathcal{R})) = \Pi C(X,\mathcal{R})$.

Therefore the product function may be considered as a bijection from $C(X,\Pi\mathcal{R})$ onto $\Pi C(X,\mathcal{R})$.

<u>Theorem 2.4.3.</u> Let X be a space and let $\mathcal{R}$ be a family of spaces.

(a) If α is a closed network on X, then $P: C_\alpha(X,\Pi\mathcal{R}) \to \Pi C_\alpha(X,\mathcal{R})$ is continuous. In addition, if α is a hereditarily closed, compact network on X, then P is a homeomorphism.

(b) If for each $R \in \mathcal{R}$, μ_R is a compatible uniformity on R and if μ is the product uniformity on $\Pi\mathcal{R}$ formed from the μ_R, then $P: C_\mu(X,\Pi\mathcal{R}) \to \Pi\{C_{\mu_R}(X,R) : R \in \mathcal{R}\}$ is a homeomorphism.

Proof. For part (a), first observe that for each $R \in \mathcal{R}$, each $A \in \alpha$ and each open V in R,

$$P^{-1}(\pi_R^{-1}([A,V])) = [A,\pi_R^{-1}(V)].$$

This establishes the continuity of P. If α is a hereditarily closed, compact network on X, then by Theorem 1.1.6, P is also a homeomorphism.

For part (b), recall that subbase for μ is given by $\{M_R^* : R \in \mathcal{R}$ and $M_R \in \mu_R\}$, where

$$M_R^* = \{(s,t) \in (\Pi\mathcal{R})\times(\Pi\mathcal{R}) : (\pi_R(s), \pi_R(t)) \in M_R\}.$$

Then one can check, for each $f \in C_\mu(X,\Pi\mathcal{R})$, $R \in \mathcal{R}$ and $M_R \in \mu_R$, that

$$P^{-1}(\pi_R^{-1}(\hat{M}_R[\pi_R \circ f])) = \hat{M}_R^*[f].$$

It follows that P is a homeomorphism. ■

A perhaps more useful type of exponential property of function spaces is the sum function. Let $\mathcal{X}$ be a family of spaces, and let $\Sigma\mathcal{X}$ denote the disjoint topological sum of the spaces in $\mathcal{X}$. For each $X \in \mathcal{X}$, let $\sigma_X: X \to \Sigma\mathcal{X}$ be the natural injection. Also if R is a space, let $R^{\mathcal{X}}$ denote the family $\{R^X : X \in \mathcal{X}\}$.

Define the <u>sum function</u>

$$S: R^{\Sigma\mathcal{X}} \to \Pi R^{\mathcal{X}}$$

by $\pi_{R^X}(S(f)) = f\circ\sigma_X$ for each $f \in R^{\Sigma\mathcal{X}}$ and $X \in \mathcal{X}$.

<u>Theorem 2.4.4.</u> The sum function $S: R^{\Sigma\mathcal{X}} \to \Pi R^{\mathcal{X}}$ is a bijection.

Proof. In this case, the goal is to show that S has as inverse the function

$$S' : \Pi R^{\mathcal{X}} \to R^{\Sigma\mathcal{X}}$$

defined by $S'(g)\circ\sigma_X = \pi_{R^X}(g)$ for each $g \in \Pi R^{\mathcal{X}}$ and $X \in \mathcal{X}$. Let $f \in R^{\Sigma\mathcal{X}}$. Then for each $X \in \mathcal{X}$, $S'(S(f))\circ\sigma_X = \pi_{R^X}(S(f)) = f\circ\sigma_X$, so that $S'\circ S(f) = S'(S(f)) = f$. On the other hand, let $g \in \Pi R^{\mathcal{X}}$. Then for each $X \in \mathcal{X}$, $\pi_{R^X}(S(S'(g))) = S'(g)\circ\sigma_X = \pi_{R^X}(g)$, so that $S\circ S'(g) = S(S'(g)) = g$. ∎

Like the product function, the sum function can be naturally restricted to continuous functions. If $\mathcal{X}$ is a family of spaces, let $C(\mathcal{X},R)$ or $C(\mathcal{X})$ denote the family $\{C(X,R) : X \in \mathcal{X}\}$.
The next theorem then follows from the fact that a function $f \in R^{\Sigma\mathcal{X}}$ is continuous if and only if $f\circ\sigma_X$ is continuous for each $X \in \mathcal{X}$.

__Theorem 2.4.5.__ If $S: R^{\Sigma\mathcal{X}} \to \Pi R^{\mathcal{X}}$ is the sum function, then $S(C(\Sigma\mathcal{X})) = \Pi C(\mathcal{X})$.

Therefore the sum function may be considered as a bijection from $C(\Sigma\mathcal{X})$ onto $\Pi C(\mathcal{X})$.

__Theorem 2.4.6.__ Let $\mathcal{X}$ be a family of spaces.

(a) If for each $X \in \mathcal{X}$, α_X is a closed network on X and if $\beta = \cup\{\sigma_X(\alpha_X) : X \in \mathcal{X}\}$, then $S: C_\beta(\Sigma\mathcal{X}) \to \Pi\{C_{\alpha_X}(X) : X \in \mathcal{X}\}$ is a homeomorphism.

(b) If μ is a compatible uniformity on R, then $S: C_\mu(\Sigma\mathcal{X}) \to \Pi C_\mu(\mathcal{X})$ is continuous. In addition, if $\mathcal{X}$ is finite, then S is a homeomorphism.

Proof. Part (a) follows from the fact that for each $X \in \mathcal{X}$, each $A \in \alpha_X$ and each open V in R,

$$S^{-1}(\pi_X^{-1}([A,V])) = [\sigma_X(A),V].$$

For part (b), let $X \in \mathcal{X}$ and let $M \in \mu$. Define $(\hat{M})^*_X = \{(f,g) \in (\Pi C_\mu(\mathcal{X}))\times(\Pi C_\mu(\mathcal{X})) : (\pi_{C(X)}(f),\pi_{C(X)}(g)) \in \hat{M}\}$, which is a member of the subbase for the product uniformity on $\Pi C_\mu(\mathcal{X})$. Then one can check that for each $f \in C_\mu(\Sigma\mathcal{X})$,

$$S(\hat{M}[f]) \subset (\hat{M}^*_X[S(f)].$$

This establishes the continuity of S. If $\mathcal{X}$ is finite, then the intersection of the sets of the form $(\hat{M})^*_X[S(f)]$, one for each $X \in \mathcal{X}$, is a neighborhood of S(f) which is contained in $S(\hat{M}[f])$. Therefore, in this case, S is a homeomorphism. ■

<u>Corollary 2.4.7.</u> If $\mathcal{X}$ is a family of spaces, then $S: C_k(\Sigma\mathcal{X}) \to \Pi C_k(\mathcal{X})$ and $S: C_p(\Sigma\mathcal{X}) \to \Pi C_p(\mathcal{X})$ are homeomorphisms.

Proof. For each $X \in \mathcal{X}$, let α_X be the family of all compact subsets of X, and let $\beta = \cup\{\sigma_X(\alpha_X) : X \in \mathcal{X}\}$. Now the set of all compact subsets of $\Sigma\mathcal{X}$ contains β and can be approximated by β. Therefore, if $\Phi: \Sigma\mathcal{X} \to \Sigma\mathcal{X}$ is the identity, then $\Phi^*: C_\beta(\Sigma\mathcal{X}) \to C_k(\Sigma\mathcal{X})$ is a homeomorphism by Theorem 2.2.7.c. A similar argument can be made using finite sets. ■

<u>5. Exponential Functions</u> If X, Y and R are any three spaces, the <u>exponential function</u>

$$E: R^{X\times Y} \to (R^Y)^X$$

is defined by $E(f)(x)(y) = f(x,y)$ for each $f \in R^{X\times Y}$, for each $x \in X$ for each $y \in Y$.

<u>Theorem 2.5.1.</u> The exponential function $E: R^{X\times Y} \to (R^Y)^X$ is a bijection.

Proof. The goal is to show that E has as inverse the function

$$E' : (R^Y)^X \to R^{X\times Y}$$

definedby $E'(g)(x,y) = g(x)(y)$ for each $g \in (R^Y)^X$ and for each $(x,y) \in X \times Y$.

First let $f \in R^{X \times Y}$ and let $(x,y) \in X \times Y$. Then $(E' \circ E(f))(x,y) = E'(E(f))(x,y) = E(f)(x)(y) = f(x,y)$, so that $E' \circ E(f) = f$. On the other hand let $g \in (R^Y)^X$, let $x \in X$ and let $y \in Y$. Then $(E \circ E'(g))(x)(y) = E(E'(g))(x)(y) = E'(g)(x,y) = g(x)(y)$, so that $E \circ E'(g) = g$. ∎

If $f \in C(X \times Y)$, then $E(f)(x) \in C(Y)$ for each $x \in X$. Therefore

$$E(C(X \times Y)) \subset (C(Y))^X.$$

If τ is any topology on $C(Y)$, then $C_\tau(Y)$ denotes this topological space. Such a topology τ is called a <u>splitting topology</u> provided that for every space X,

$$E(C(X \times Y)) \subset C(X, C_\tau(Y)).$$

Note that any topology smaller than a splitting topology is a splitting topology.

<u>Theorem 2.5.2.</u> The compact-open topology is always a splitting topology.

Proof. Let X and Y be spaces, and let $f \in C(X \times Y)$. To show that $E(f)$ is continuous, let $x \in X$ and let $[B,W]$ be a subbasic neighborhood of $E(f)(x)$ in $C_k(Y)$. For each $y \in B$ there exists a neighborhood U_y of x in X and a neighborhood V_y of y in Y such that $f(U_y \times V_y) \subset W$. Since B is compact, there exist $y_1, \ldots, y_n \in B$ such that $B \subset V_{y_1} \cup \ldots \cup V_{y_n}$. If $U = U_{y_1} \cap \ldots \cap U_{y_n}$, then U is a neighborhood of x such that $E(f)(U) \subset [B,W]$. ∎

It follows from Theorem 2.5.2 that if β is any compact network on Y, then $C_\beta(Y)$ has a splitting topology.

A topology τ on $C(Y)$ is called a <u>conjoining topology</u> provided that for every space X,

$$C(X, C_\tau(Y)) \subset E(C(X \times Y)).$$

A conjoining topology is sometimes called an admissible topology and sometimes called a jointly continuous topology. Any topology larger than a conjoining topology is a conjoining topology.

Theorem 2.5.3. A topology τ on C(Y) is a conjoining topology if and only if the evaluation function e: $C_\tau(Y)\times Y \to R$ is continuous.

Proof. Suppose that e is continuous. Let X be a space and let $g \in C(X,C_\tau(Y))$. It suffices to show that

$$E^{-1}(g) = e\circ(g\times id),$$

where id is the identity map on Y. If $(x,y) \in X\times Y$, then $(e\circ(g\times id))(x,y) = e(g(x),y) = g(x)(y) = E^{-1}(g)(x,y)$.

On the other hand, suppose that τ is conjoining. Then take $X = C_\tau(Y)$. It follows that $e = e\circ(id\times id) = E^{-1}(id)$, which is continuous. ■

Corollary 2.5.4. Let Y and R be spaces.

(a) If Y is locally compact, then the compact-open topology on C(Y) is a conjoining topology.

(b) If μ is a compatible uniformity on R, then the topology of uniform convergence on C(Y) is a conjoining topology.

Corollary 2.5.5. If Y is locally compact, then for any space X, the exponential function E is a bijection from $C(X\times Y)$ onto $C(X,C_k(Y))$.

The next theorem establishes the continuity properties of E. If α is a closed network on X and β is a closed network on Y, then define

$$\alpha\times\beta = \{A\times B : A \in \alpha \text{ and } B \in \beta\},$$

which is a closed network on $X\times Y$. Also $\alpha\times k$ denotes $\{A\times B : A \in \alpha$ and B is compact in Y$\}$.

Theorem 2.5.6. Let Y be a locally compact space and let X be any space.

(a) If α is a hereditarily closed, compact network on X, then E: $C_{\alpha\times k}(X\times Y) \to C_\alpha(X,C_k(Y))$ is a homeomorphism.

(b) If Y is compact, if μ is a compatible uniformity on R and if ν is the uniformity on $C_\mu(Y)$ induced by μ, then E: $C_\mu(X\times Y) \to C_\nu(X,C_\mu(Y))$ is a homeomorphism.

Proof. Part (a) follows from Theorem 1.1.6, Corollary 2.2.5 and the easily shown fact that

$$E([A\times B,W]) = [A,[B,W]]$$

for each $A \in \alpha$, each compact B in Y and each open W in R.

For part (b), it suffices to show for each $f \in C_\mu(X\times Y)$ and each $M \in \mu$, that

$$E(\hat{M}[f]) = \hat{\hat{M}}[E(f)].$$

Let $g \in \hat{M}[f]$, let $x \in X$ and let $y \in Y$. Then $(f(x,y),g(x,y)) \in M$, so that $(E(f)(x)(y),E(g)(x)(y)) \in M$. Since y is arbitrary, then $(E(f)(x),E(g)(x)) \in \hat{M}$. Also since x is arbitrary, $(E(f),E(G)) \in \hat{\hat{M}}$, so that $E(g) \in \hat{\hat{M}}[E(f)]$. For the other inclusion, let $g \in \hat{\hat{M}}[E(f)]$ and let $(x,y) \in X\times Y$. Then $(E(f)(x),g(x)) \in \hat{M}$, so that $(f(x,y),E^{-1}(g)(x,y)) = (E(f)(x)(y)),g(x)(y)) \in M$. Therefore $E^{-1}(g) \in \hat{M}[f]$, and thus $g \in E(\hat{M}[f])$. ∎

<u>Corollary 2.5.7.</u> If Y is locally compact, then for each space X, E: $C_k(X\times Y) \to C_k(X,C_k(Y))$ is a homeomorphism.

Proof. Let $\beta = \{A\times B : A$ is compact in X and B is compact in Y$\}$. Because of Theorem 2.2.7.c, it is enough to show that the family of all compact subsets of X×Y can be approximated by β. So let C be a compact subset of X×Y and let W be an open subset of X×Y containing C. Now take A and B to be the projections of C into X and Y, respectively. For each $z \in C$, there exist open sets U_z in X and V_z in Y such that

$z \in U_z \times V_z$ and $(\overline{U_z} \cap A) \times (\overline{V_z} \cap B) \subset W$. Then for each $z \in C$, define $A_z = \overline{B_z} \cap A$ and $B_z = \hat{V}_z \cap B$. Since C is compact, there exist $z_1, \ldots, z_n \in C$ such that $C \subset (U_{z_1} \times V_{z_1}) \cup \ldots \cup (U_{z_n} \times V_{z_n})$. Therefore $C \subset (A_{z_1} \times B_{z_1}) \cup \ldots \cup (A_{z_n} \times B_{z_n}) \subset W$. ∎

The local compactness in Theorem 2.5.6.a and Corollary 2.5.7 is only needed to insure that E is onto. This may also be obtained by taking X×Y as a k-space, as given by the next corollary.

Corollary 2.5.8. If X×Y is a k-space, then $E: C_k(X \times Y) \to C_k(X, C_k(Y))$ is a homeomorphism.

Proof. To show that E is onto, let $g \in C_k(X, C_k(Y))$. Since X×Y is a k-space, it suffices to show that $E^{-1}(g)|_{A \times B}$ is continuous, where A and B are compact subsets of X and Y, respectively. Let $j: B \to Y$ be the inclusion map, so that the induced function $j^*: C_k(Y) \to C_k(B)$ is continuous. Also the evaluation function $e: C_k(B) \times B \to R$ is continuous since B is compact. Now it is easy to check that $E^{-1}(g)|_{A \times B} = e \circ (j^* \times id) \circ (g|_A \times id)$, where $id: B \to B$ is the identity map. ∎

In this section, no use has been made of any particular topological property of the range space R. For this reason, all the theorems and corollaries in this section are true for any topolgoical space R. Therefore the exponential function can be applied to prove a useful result known as the Whitehead Theorem.

Lemma 2.5.9. If $f: X \to Y$ is a continuous surjection, then f is a quotient map if and only if for every topological space R and every function $g: Y \to R$, the continuity of $g \circ f$ implies the continuity of g.

Proof. Let $f: X \to Y$ be a quotient map, let $g: Y \to R$ be a function with $g \circ f$ continuous, and let W be open in R. Then $f^{-1}(g^{-1}(W)) = (g \circ f)^{-1}(W)$ is open in X.

Since f is a quotient map, $g^{-1}(W)$ is open in Y. This establishes the continuity of g.

Conversely, to see that f is a quotient map, let V be a subset of Y with $f^{-1}(V)$ open in X. Define R to be the set Y with the quotient topology induced by f (i.e., the largest topology making f continuous), and let $g: Y \to R$ be the identity map. Now $g \circ f$ is a quotient map, so that since $(g \circ f)^{-1}(g(V)) = f^{-1}(V)$, then g(V) is open in R. But since g is continuous, $V = g^{-1}(g(V))$ is open in Y. Therefore f is a quotient map. ∎

<u>Theorem 2.5.10.</u> If Y is locally compact and $q: X \to Z$ is a quotient map, then $q \times id: X \times Y \to Z \times Y$ is a quotient map.

Proof. By Lemma 2.5.9, it suffices to prove that if R is a topological space and $g: Z \times Y \to R$ is a function such that $g \circ (q \times id)$ is continuous, then g is continuous. Let $f = g \circ (q \times id)$, and let $E: C_k(X \times Y, R) \to C_k(X, C_k(Y,R))$ and $E': C_k(Z \times Y, R) \to C_k(Z, C_k(Y,R))$ be the exponential functions, which are homeomorphisms by Corollary 2.5.7. Now E' is actually defined on $R^{Z \times Y}$, so that $E'(g)$ is a function from Z to R^Y. To show that $E'(g) \circ q = E(f)$, let $x \in X$ and $y \in Y$. Then $(E'(g) \circ q)(x)(y) = E'(g)(q(x))(y) = g(q(x),y) = (g \circ (q \times id))(x,y) = f(x,y) = E(f)(x)(y)$. Since E(f) is continuous, it follows from the other half of Lemma 2.5.9 that $E'(g)$ is continuous. Therefore g is continuous as desired. ∎

Since the class of k-spaces is precisely the class of all quotient images of locally compact spaces, and since the product of two locally compact spaces is locally compact, then Theorem 2.5.10 has the following corollary.

<u>Corollary 2.5.11.</u> The product of a k-space and a locally compact space is a k-space.

<u>6. Exercises and Problems for Chapter II.</u>

1. If α is a hereditarily closed, compact network on X and ν is a compatible uniformity on R, then Φ: $C_\alpha(X,Y) \times C_\nu(Y,R) \to C_{\alpha,\nu}(X,R)$ is continuous.

2. Let $f \in C(X,Y)$ and let α and β be compact networks on X and Y, respectively.

(a) Then $f^*: C_\beta(Y,\mathbb{R}) \to C_\alpha(X,\mathbb{R})$ is continuous if and only if $f(\alpha)$ can be approximated by β.

(b) Also $f^*: C_\beta(Y,\mathbb{R}) \to C_\alpha(X,\mathbb{R})$ is open onto its image if and only if f(X) is closed in Y and $\{B \cap f(X) : B \in \beta\}$ can be approximated by $f(\alpha)$.

3. Let Δ: $X \to C_k(C_k(X))$ be the diagonal function defined by $\Delta(x) = e_x$ for each $x \in X$.

(a) If (x_i) is a net (if $\mathcal{F}$ is filter base, resp.) on X such that $\Delta((x_i))$ ($\Delta(\mathcal{F})$, resp.) has a cluster point in $C_k(C_k(X))$, then (x_i) ($\mathcal{F}$,resp.) has a cluster point in X.

(b) Conclude that Δ is an open function onto its image, and that whenever it is continuous then it is a closed embedding.

4. Let $\mathcal{X}$ be a family of spaces, and let μ be a compatible uniformity on R. Then the sum function S: $C_\mu(\Sigma\mathcal{X}) \to \Pi C_\mu(\mathcal{X})$ is a homeomorphism if $\Pi C_\mu(\mathcal{X})$ has the "box product topology".

5. Let Y be a paracompact, and let β be a closed network on Y. Then the following are equivalent.

(a) $C_\beta(Y)$ has a splitting topology.

(b) β is a compact network on Y.

(c) For all X and all $B \in \beta$, the projection map π_X: $X \times B \to X$ is closed.

6. Fine Topology (cf. Exercise I.3.1). Let (R,ρ) be a metric space.

(a) If f: $X \to Y$ is a perfect map, then the induced function f^*: $C_{f_\rho}(Y) \to$

$C_{f_\rho}(X)$ is a closed embedding.

(b) If $\mathcal{X}$ is a family of spaces, then the sum function S: $C_{f_\rho}(\Sigma\mathcal{X}) \to \Pi C_{f_\rho}(\mathcal{X})$ is continuous. In addition, if $\mathcal{X}$ is finite, then S is a homeomorphism.

(c) Let Y be compact, and let d be the supremum metric $\hat{\rho}$ on C(Y). Then for any space X, the exponential function E: $C_{f_\rho}(X\times Y) \to C_{f_d}(X,C_{f_\rho}(Y))$ is a continuous bijection.

7. Graph Topology (cf. Exercise I.3.2).

(a) If X is compact, then for any Y and R, the composition function Φ: $C_\gamma(X,Y) \times C_\gamma(Y,R) \to C_\gamma(X,R)$ is continuous.

(b) For any $g \in C(Y,R)$, the induced function g_*: $C_\gamma(X,Y) \to C_\gamma(X,R)$ is continuous. If g is an embedding, so is g_*.

(c) Let S be the Sorgenfrey line ($\mathbb{R}$ with the topology generated by intervals of the form [a,b)) and let F: S $\to \mathbb{R}$ be the identity map. Then f^*: $C_\gamma(\mathbb{R}) \to C_\gamma(S)$ is not continuous.

(d) If $\mathcal{R}$ is a family of spaces, then the product function P: $C_\gamma(X,\Pi\mathcal{R}) \to \Pi C_\gamma(X,\mathcal{R})$ is continuous. In addition, if $\mathcal{R}$ is finite, then P is a homeomorphism.

(e) If $\mathcal{X}$ is a family of spaces, then the sum function S: $C_\gamma(\Sigma\mathcal{X}) \to \Pi C_\gamma(\mathcal{X})$ is continuous. In addition if $\mathcal{X}$ is finite, then S is a homeomorphism.

8. Direct and Inverse Limits. Let $\{R_i;\psi_i^j\}$ be an inverse system of topological spaces and continuous functions, and let $\{X_i;\phi_i^j\}$ be a direct system of topological spaces and continuous functions.

(a) If α is a closed network on X, then

$$P\circ q_*: C_\alpha(X,\varprojlim\{R_i;\psi_i^j\}) \to \varprojlim\{C_\alpha(X,R_i);(\psi_i^j)_*\}$$

is a continuous injection, where q is the inclusion map of $\varprojlim\{R_i;\psi_i^j\}$ into the product of the R_i, and P is the product function. If in addition, α is a hereditarily closed, compact network on X, then $P\circ q_*$ is a homeomorphism.

(b) If for each i, μ_i is a compatible uniformity on R_i, and if each ψ_i^j is uniformly continuous, then $\{C_{\mu_i}(X,R_i);(\psi_i^j)_*\}$ is an inverse system. If μ is the product uniformity restricted to $\varprojlim\{R_i;\psi_i^j\}$ which is formed from the μ_i, then

$$P\circ q_*\colon C_\mu(X,\varprojlim\{R_i;\psi_i^j\}) \to \varprojlim\{C_{\mu_i}(X,R_i);(\psi_i^j)_*\}$$

is a homeomorphism.

(c) If for each i, α_i is a closed network on X_i such that $\phi_i^j(\alpha_i) \subset \alpha_j$ whenever $i < j$, and if β is a closed network on $\varinjlim\{X_i;\phi_i^j\}$ which can be approximated by $p(\cup_i\alpha_i)$, where $p\colon \Sigma_i X_i \to \varinjlim\{X_i;\phi_i^j\}$ is the natural projection, then

$$S\circ p\colon C_\beta(\varinjlim\{X_i;\phi_i^j\}) \to \varprojlim\{C_{\alpha_i}(X_i);(\phi_i^j)^*\}$$

is a homeomorphism.

(d) If μ is a compatible uniformity on R, then

$S\circ p^*\colon C_\mu(\varinjlim\{X_i;\phi_i^j\}) \to \varprojlim\{C_\mu(X_i);(\phi_i^j)_*\}$ is a continuous bijection.

(e) If $\{X_i;\phi_i^j\}$ is a direct system, then $C_p(\varinjlim\{X_i;\phi_i^j\})$ is homeomorphic to $\varprojlim\{C_p(X_i);(\phi_i^j)^*\}$.

(f) If X is a k-space and if $\{X_i;\phi_i^j\}$ is the direct system of compact subspaces of X, then $C_k(X)$ is homeomorphic to $\varprojlim\{C_k(X_i);(\phi_i^j)^*\}$.

Chapter III

CONVERGENCE AND COMPACT SUBSETS

This chapter contains some classical results about function spaces, including a version of the Ascoli Theorem.

1. Convergence. Convergence in C(X) can be studied and related to convergence in X and R by either using nets or using filters. Nets will be used in this section, but every statement made using nets has its analog in terms of filters (or filter bases).

Let (f_i) be a net in C(X) (i.e., a function from a directed set I into C(X)), and let $f \in C(X)$. If (f_i) converges to f in the space $C_p(X)$, then (f_i) is said to converge pointwise to f. Also if μ is a compatible uniformity on R and (f_i) converges to f in the space $C_\mu(X)$, then (f_i) is said to converge uniformly to f (with respect to μ). Furthermore, if α is a closed network on X, then (f_i) converges uniformly to f on α (with respect to μ) if (f_i) converges to f in $C_{\alpha,\mu}(X)$. So by Theorem 1.2.3, convergence in $C_k(X)$ is precisely uniform convergence on compact sets.

Theorem 3.1.1. If (f_i) is a net in C(X) and $f \in C(X)$, then (f_i) converges pointwise to f if and only if for every $x \in X$, $(f_i(x))$ converges to f(x) in R.

Proof. Suppose (f_i) converges pointwise to f, and let $x \in X$ and let V be a neighborhood of f(x) in R. Then $f \in [x,V]$, so that there is an index i_0 such that for every $i \geq i_0$, $f_i \in [x,V]$. Therefore for every $i \geq i_0$, $f_i(x) \in V$, so that $(f_i(x))$ converges to f(x) in R.

For the converse, let $(f_i(x))$ converge to f(x) in R for every $x \in X$, and let $f \in [x_1,V_1] \cap \dots \cap [x_n,V_n]$. Then for each integer j between 1 and n, there exists an index i_j such that for every $i \geq i_j$, $f_i(x_j) \in V_j$. Let i_0 be an index greater than or equal to each i_j. Then if $i \geq i_0$, $f_i \in [x_1,V_1] \cap \dots \cap [x_n,V_n]$, so that (f_i) converges pointwise to f. ∎

There is another natural kind of convergence called continuous convergence defined as follows. Let $(f_i)_{i\in I}$ be a net in C(X), and let $(x_j)_{j\in J}$ be a net in X. Direct the set I×J by: $(i_1,j_1) \leq (i_2,j_2)$ if and only if $i_1 \leq i_2$ and $j_1 \leq j_2$. Then $(f_i(x_j))_{(i,j)\in I\times J}$ is a net in R, which will be denoted by $(f_i(x_j))$. The net (f_i) <u>converges continuously</u> to f provided that for every $x \in X$ and every net (x_j) in X converging to x in X, the net $(f_i(x_j))$ converges to f(x) in R. It turns out that continuous convergence is related to the splitting topology and conjoining topology concepts which were introduced in section 2.5.

<u>Theorem 3.1.2.</u> A topology τ on C(X) is a splitting topology if and only if whenever (f_i) is a net in C(X) converging continuously to $f \in C(X)$ then (f_i) converges to f in $C_\tau(X)$.

Proof. First suppose that τ is a splitting topology and that $(f_i)_{i\in I}$ is a net in C(X) converging continuously to $f \in C(X)$. Define the space $\tilde{I}$ to be the set $I \cup \{\tilde{i}\}$, where $\tilde{i}$ is an element not in I, along with the following topology. Each singleton set {i} is open for $i \in I$, and basic neighborhoods of $\tilde{i}$ have the form $\{i \in I : i \geq i_0\} \cup \{\tilde{i}\}$ for some $i_0 \in I$. Then define Φ: $\tilde{I}\times X \to R$ by $\Phi(i,x) = f_i(x)$ for each $i \in I$ and $x \in X$, and $\Phi(\tilde{i},x) = f(x)$ for each $x \in X$. The fact that (f_i) converges continuously to f ensures that Φ is continuous. Therefore, since τ is splitting, then $E(\Phi) \in C(\tilde{I},C_\tau(X))$, where E is the exponential function. Since $(i)_{i\in I}$ is a net in $\tilde{I}$ converging to $\tilde{i}$, then $(f_i) = (E(\Phi)(i))$ converges to $E(\Phi)(\tilde{i}) = f$ in $C_\tau(X)$.

For the converse, suppose that continuous convergence implies convergence in $C_\tau(X)$. To see that τ is a splitting topology, let Z be any space and let $\Phi \in C(Z\times X)$. To show that $E(\Phi) \in C(Z,C_\tau(X))$, where E is the exponential function, let (z_i) be a net in Z converging to $z \in Z$. Define $f = E(\Phi)(z)$, and for each i, define $f_i = E(\Phi)(z_i)$. For each net (x_j) in X converging to $x \in X$, $(f_i(x_j)) = (\Phi(z_i,x_j))$ converges to $\Phi(z,x) = f(x)$ in R, so that (f_i) converges continuously to f. Therefore $(E(\Phi)(z_i)) = (f_i)$ converges to $f = E(\Phi)(z)$ in $C_\tau(X)$, so that $E(\Phi)$ is continuous. ∎

The dual result concerning conjoining topologies is easier to prove because of the relationship with the evaluation function given by Theorem 2.5.3.

<u>Theoerem 3.1.3.</u> A topology τ on C(X) is a conjoining topology if and only if whenever (f_i) is a net in C(X) converging to f in $C_\tau(X)$ then (f_i) converges continuously to f.

Proof. Suppose that τ is a conjoining topology and that (f_i) converges to f in $C_\tau(X)$. Let (x_j) be a net in X converging to $x \in X$. Then the net $((f_i,x_j))$ converges to (f,x) in $C_\tau(X)\times X$. Since τ is conjoining, the evaluation function $e: C_\tau(X)\times X \to R$ is continuous. Thus $(f_i(x_j)) = (e(f_i,x_j))$ converges to $e(f,x) = f(x)$ in R, so that (f_i) converges continuously to f.

Conversely, suppose that convergence in $C_\tau(X)$ implies continuous convergence. It suffices to show that the evaluation function $e: C_\tau(X)\times X \to R$ is continuous. Let $((f_i,x_i))$ be a net in $C_\tau(X)\times X$ converging to (f,x). Then (f_i) converges to f in $C_\tau(X)$ and (x_i) converges to x in X. Since (f_i) converges continuously, then $(e(f_i,x_i)) = (f_i(x_i))$ converges to $f(x) = e(f,x)$ in R, so that e is continuous. ∎

<u>Corollary 3.1.4.</u> Let (f_i) be a net in C(X) and let $f \in C(X)$.

(a) If α is a compact network on X and (f_i) converges continuously to f, then (f_i)

converges to f in $C_\alpha(X)$.

(b) If α is a closed neighborhood base on X and (f_i) converges to f in $C_\alpha(X)$, then (f_i) converges continuously to f.

Corollary 3.1.5. Let X be a locally compact space, let (f_i) be a net in C(X), and let $f \in C(X)$. Then (f_i) converges to f in $C_k(X)$ if and only if (f_i) converges continuously to f.

Corollary 3.1.6. If σ is any splitting topology on C(X) and τ is any conjoining topology on C(X), then $C_\sigma(X) \le C_\tau(X)$.

So there is at most one topology on C(X) which is both splitting and conjoining. Such a topology would be the maximum splitting topology and the minimum conjoining topology. This topology is the compact-open topology whenever X is locally compact.

Theorem 3.1.7. Let τ be any topology on C(X).

(a) If $C_\tau(X) \le C_k(X)$, then τ is a splitting topology.

(b) If X is locally compact and τ is a splitting topology, then $C_\tau(X) \le C_k(X)$.

(c) If X is locally compact and $C_k(X) \le C_\tau(X)$, then τ is a conjoining topology.

(d) If τ is a conjoining topology, then $C_k(X) \le C_\tau(X)$.

Proof. Parts (a) through (c) follow from previous results. To establish part (d), suppose that the evaluation function e: $C_\tau(X)\times X \to \mathbb{R}$ is continuous, and let [A,V] be a subbasic open set in $C_k(X)$. To see that $[A,V] \in \tau$, let $f \in [A,V]$. For each $x \in A$, there exists a neighborhood U_x of x in X and a neighborhood W_x of f in $C_\tau(X)$ such that $e(W_x\times U_x) \subset V$. Since A is compact, there exist $x_1,\dots,x_n \in A$ such that $A \subset U_{x_1} \cup \dots \cup U_{x_n}$. Then $W_{x_1} \cap \dots \cap W_{x_n}$ is a neighborhood of f in τ which is contained in [A,V]. ∎

The local compactness in part (b) of Theorem 3.1.7 may be dropped if τ is a set-open topology (cf. Exercise 1(d)); although this is not true for general τ. The local compactness in part (c) is needed even when τ is a set-open topology.

The topology on C(X) generated by the union of all the splitting topologies on C(X) is the unique maximum splitting topolgoy on C(X). On the other hand, there need not be a minimum conjoining topology on C(X). In fact C(X) has a minmum conjoining topology if and only if X is locally compact (cf. Exercise 1(b)), in which case this topology is the compact-open topology.

2. Compact subsets. This section covers a version of the Ascoli Theorem, which gives a characterization of the compact subsets of $C_k(X)$.

If X and R are spaces, a subset F of R^X is evenly continuous provided that for each $x \in X$, each $r \in R$ and each neighborhood V of r, there exists a neighborhood U of x and a neighborhood V′ of r such that $f(U) \subset V$ for every $f \in F$ satisfying $f(x) \in V'$. Observe that every member of an evenly continuous family is automatically continuous.

Lemma 3.2.1. If F is an evenly continuous subset of C(X), then the closure of F in R^X is evenly continuous, and is therefore equal to the closure of F in $C_p(X)$.

Proof. Let $\overline{F}$ be the closure of F in R^X. Let $x \in X$, let $r \in R$, and let V be a closed neighborhood of r. Since R is evenly continuous, there exist neighborhoods U of x and V′ of r such that $f(U) \subset V$ whenever $f \in F$ with $f(x) \in V'$. Now let $f \in \overline{F}$ with $f(x) \in V'$. Then there exists a net (f_i) in F which converges pointwise to f. So for some index i_0, $f_i(x) \in V'$ whenever $i \geq i_0$. If $u \in U$, then $f_i(u) \in V$ whenever $i \geq i_0$, so that $f(u) \in V$ since V is closed. Therefore $f(U) \subset V$, and thus $\overline{F}$ is evenly continuous. ∎

Lemma 3.2.2 If Z is a compact space and E: $C(Z\times X) \to (C(X))^Z$ is the exponential function, then for each $f \in C(Z\times X)$, $E(f)(Z)$ is an evenly continuous subset of C(X).

Proof. Let $x \in X$, let $r \in R$ and let V be a closed neighborhood of r. If π_Z: $Z\times X \to Z$ is the projection map, define $Y = \pi_Z((Z\times\{x\}) \cap f^{-1}(V))$, which is compact. Then the projection map π_X: $Y\times X \to X$ is closed, so that the set $U = X\setminus\pi_X((Y\times X)\setminus f^{-1}(V))$ is open in X. If $g \in E(f)(Z)$ with $g(x) \in V$, then $g \in E(F)(Y)$, so that $g(U) \subset V$. ∎

Lemma 3.2.3. If F is an evenly continuous subspace of $C_p(X)$, then the evaluation function e: $F\times X \to R$ is continuous. Conversely, if F is a compact subspace of $C_p(X)$ with e: $F\times X \to R$ continuous, then F is evenly continuous.

Proof. Suppose F is evenly continuous, and let $(f,x) \in F\times X$ and let V be a neighborhood of $e(f,x) = f(x)$ in R. Then there is a neighborhood U of x and a neighborhood V' of f(x) such that $g(U) \subset V$ whenever $g \in F$ with $g(x) \in V'$. But then $e([x,V']\times U) \subset V$, making e continuous. The converse follows from Lemma 3.2.2 since $E(e)(f) = F$, where E: $C(F\times X) \to (C(X))^F$ is the exponential map. ∎

A subset F of C(X) is pointwise bounded at $x \in X$ provided that the closure of $e_x(F)$ is compact, where e_x: $C_p(X) \to R$ is the evaluation function at x. Recall that $e_x(F) = \{f(x) : f \in F\}$, which is also sometimes denoted by F(x). Then F is pointwise bounded provided it is pointwise bounded at every point of X. Since e_x is continuous, every compact subset of $C_p(X)$ is pointwise bounded.

A topology τ on C(X) is called proper if it is finer than or equal to the topology of pointwise convergence. Then for a proper topology τ, a compact subset of $C_\tau(X)$ is closed and pointwise bounded. Under certain conditions a compact subset of $C_\tau(X)$ is also evenly continuous. A topology τ on C(X) is called a hyper-Ascoli topology provided that every compact subset of $C_\tau(X)$ is closed, pointwise bounded, and evenly continuous. On the other hand, a topology τ on $C(X)$ is called a <u>hypo-Ascoli topology</u> provided that every closed, pointwise bounded, and evenly continuous subset of $C_\tau(X)$ is compact. If τ is both hypo-Ascoli and hyper-Ascoli then it is called an <u>Ascoli topology</u>.

These concepts are related to splitting and conjoining topologies. A topology τ on $C(X)$ is called a <u>weakly splitting topology</u> provided that for every compact space Z,

$$E(C(Z\times X)) \subset C(Z, C_\tau(X)),$$

where $E\colon R^{Z\times X} \to (R^X)^Z$ is the exponential function. Also a topology τ on $C(X)$ is called a <u>weakly conjoining topology</u> provided for every compact space Z,

$$C(Z, C_\tau(X)) \subset E(C(Z\times X)).$$

These concepts are related by the following fundamental theorem.

<u>Theorem 3.2.4.</u> Let τ be a proper topology on $C(X)$.

(a) Then τ is a hypo-Ascoli topology if and only if it is a weakly splitting topology.

(b) Also τ is a hyper-Ascoli topology if and only if it is a weakly conjoining topology.

Proof. First suppose that τ is a weakly splitting topology, and let F be a closed, pointwise bounded, and evenly continuous subset of $C_\tau(X)$. By Lemma 3.2.1, the closure Z of F in $C_p(X)$ is evenly continuous, so that the evaluation function $e\colon Z\times X \to R$ is continuous by Lemma 3.2.3. Since Z is a closed subet of $\Pi\{e_x(F) : x \in X\}$, which is compact, then Z is compact. Also since τ is a weakly splitting topology, $E(e)$ is continuous, where $E\colon C(Z\times X) \to (C_\tau(X))^Z$ is the exponential function. Therefore $E(e)(Z)$ is a compact subset of $C_\tau(X)$. But $E(e)$ is just the inclusion map from Z into $C_\tau(X)$, so that Z is a compact subspace of $C_\tau(X)$. Then since F is a closed subset of Z, F is compact, and thus τ is a hypo-Ascoli topology.

Next suppose that τ is a weakly conjoining topology, and let F be a compact subset of $C_\tau(X)$. If e: $F\times X \to R$ is the evaluation function, then E(e) is the inclusion map from F into $C_\tau(X)$, where E: $C(F\times X) \to (C_\tau(X))^F$ is the exponential function. Since τ is weakly conjoining, then e is continuous. So by Lemma 3.2.2, F must be evenly continuous, and τ is a hyper-Ascoli topology.

Now suppose that τ is a hyper-Ascoli topology, and let Z be a compact space and let $f \in R^{Z\times X}$ be such that $E(f) \in C(Z,C_\tau(X))$, where E: $R^{Z\times X} \to (R^X)^Z$ is the evaluation function. Then E(f)(Z) is a compact subset of $C_\tau(X)$, and is hence evenly continuous. By Lemma 3.2.3, the evaluation function e: $E(f)(Z) \to R$ is continuous. But $E(e\circ(E(F)\times id)) = E(f)$, so that $f = e\circ(E(f)\times id)$ is continuous. Therefore τ is a weakly conjoining topology.

Finally suppose that τ is a hypo-Ascoli topology, and let Z be a compact space and let $f \in C(Z\times X)$. Since $E_p(f)$ is continuous, where E_p: $C(Z\times X) \to (C_p(X))^Z$ is the exponential function, then $E_p(f)(Z)$ is a compact subset of $C_p(X)$. IF E_τ: $C(Z\times X) \to (C_\tau(X))^Z$ is also the exponential function, then since τ is proper, $E_\tau(f)(Z)$ is at least a closed and pointwise bounded subset of $C_\tau(X)$. Also Lemma 3.2.2 shows that $E_\tau(f)(Z)$ is evenly continuous, so that since τ is hypo-Ascoli, $E_\tau(f)(Z)$ is compact. A compact (Hausdorff) space is minimal Hausdorff, so that $E_\tau(f)(Z) = E_p(f)(Z)$. Therefore $E_\tau(f)$ is continuous, and hence τ is a weakly splitting topology. ■

The compact-open topology is always a splitting topology, so that it is a hypo-Ascoli topology. Whenever X is locally compact, $C_k(X)$ also has a conjoining topology, which is therefore an Ascoli topology. The next lemma gives a more general condition which insures that the compact-open topology is weakly conjoining.

<u>Lemma 3.2.5.</u> If X is a k-space, then $C_k(X)$ has a weakly conjoining topology.

Proof. If Z is a compact space, then $Z\times X$ is a k-space by Corollary 2.5.11. Therefore the proof of Corollary 2.5.8 shows that the exponential function E: $C(Z\times X) \to C(Z,C_k(X))$ is onto, so that $C_k(X)$ has a weakly conjoining topology. ■

The next theorem, known as the Ascoli Theorem or the Arzela–Ascoli Theorem, now follows from Lemma 3.2.5 and Theorem 3.2.4.

Theorem 3.2.6. If X is a k-space, then a subset of $C_k(X)$ is compact if and only if it is closed, pointwise bounded, and evenly continuous.

The "evenly continuous" property in Theorem 3.2.6 may be replaced by the property of "equicontinuous" (cf Exercise 4), which is a property dependent upon a compatible uniform structure on R.

Countably compact and sequentially compact subsets of $C_k(X)$ and $C_p(X)$ are studied in Chapter 5.

3. Exercises and Problems for Chapter III.

1. Conjoining and Splitting Topologies. (Arens and Dugundji [1951])

(a) Let $p \in C([0,1],R)$ be a path such that $p(0) \neq p(1)$, let $V = R \setminus \{p(1)\}$, and let τ be a topology on C(X,R) which is smaller than or equal to every conjoining topology on C(X,R). If A is a closed subset of X such that [A,V] contains a nonempty open subset of $C_\tau(X,R)$, then A is compact.

(b) The function space C(X) has a minimum conjoining topology if and only if X is locally compact.

(c) The compact-open topology on C(X) is conjoining if and only if X is locally compact.

(d) If α is a closed network on X, then $C_\alpha(X)$ is a splitting topology if and only if $C_\alpha(X) \leq C_k(X)$.

2. (Noble [1969]) The following are equivalent for a space X.

(a) The compact-open topology on C(X) is an Ascoli topology.

(b) The compact-open topology on $C(X)$ is conjoining.

(c) The compact-open topology on $C(X)$ is weakly conjoining.

(d) For each locally compact space Z, $C(Z,C_k(X)) \subset E(C(Z\times X))$.

(e) The diagonal function $\Delta\colon X \to C_k(C_k(X))$ is continuous.

3. Graph Topology. (cf. Exercises I.3.2 and II.6.7)

(a) A sequence (f_n) in $C(X)$ is <u>eventually supported on a compact set</u> provided there exists a compact subset A of X and an integer m such that for every $n \geq m$, $f_n|_{X\setminus A} = f_m|_{X\setminus A}$. Let X be a paracompact, locally compact space. Then a sequence (f_n) in $C(X)$ converges to f in $C_\gamma(X)$ if and only if (f_n) converges continuously to f and is eventually supported on a compact set.

(b) A subset F of $C(X)$ is <u>supported on a compact set</u> provided there exists a compact subset A of X such that for every $f,g \in F$, $f|_{X\setminus A} = g|_{X\setminus A}$. Also F is <u>almost supported on a compact set</u> provided that it is the union of finitely many subsets of $C(X)$ each supported on a compact set. If X is a paracompact, locally compact space, then a subset of $C_\gamma(X)$ is compact if and only if it is closed, pointwise bounded, evenly continuous, and almost supported on a compact set.

4. Equicontinuity. (Kelley [1955]) Let $F \subset C(X)$, and let $x \in X$. If μ is any compatible uniformity on $\mathbb{R}$, then F is <u>equicontinuous at</u> x (with respect to μ) provided that for every $M \in \mu$, there exists a neighborhood U of x such that $f(U) \subset M[f(x)]$ whenever $f \in F$. Furthermore, F is <u>equicontinuous</u> (with respect to μ) provided that it is equicontinuous at every $x \in X$. As a special case, a subset $F \subset C(X,\mathbb{R})$ is equicontinuous if and only if for every $x \in X$ and $\varepsilon > 0$, there exists a neighborhood U of x such that for every $f \in F$ and $u \in U$, $|f(u) - f(x)| < \varepsilon$.

(a) If F is equicontinuous with respect to some compatible uniformity on $\mathbb{R}$, then F is evenly continuous.

(b) If F is evenly continuous and pointwise bounded at $x \in X$, then F is equicontinuous at x with respect to every compatible uniformity on $\mathbb{R}$.

(c) Let X be pseudocompact, let R be metrizable, and let μ be a compatible uniformity on R. Then a subset of $C_\mu(X)$ is compact if and only if it is closed, pointwise bounded, and equicontinuous.

(d) If F is an equicontinuous subset of C(X,R), then for any $n \in \omega$, the set $\{f_1+...+f_k : k=1,...,n \text{ and } f_1,...,f_k \in F\}$ is equicontinuous.

(e) If F is an equicontinuous subset of C(X,R) such that $f(X) \subset [-1,1]$ for each $f \in F$, then for any $n \in \omega$, $\{f_1...f_k : k=1, ..., n \text{ and } f_1,...,f_k \in F\}$ is equicontinuous.

5. Simple Even Continuity. (Noble [1969]) A subset F of C(X) is called <u>simply evenly continuous</u> provided that for every $x \in X$, every $r \in R$, every neighborhood V of r, and every net (f_i) in F, there exist neighborhoods U of x and V′ of r such that for each $u \in U$ there is a subset (f_{i_j}) of (f_i) having the property that either $(f_{i_j}(u))$ is eventually in V or $(f_{i_j}(u))$ is not eventually in V′. Then a subset of $C_p(X)$ is compact if and only if it is closed, pointwise bounded, and simply evenly continuous.

6. Isbell Topology. (Isbell [1975]) Let $\Omega(X)$ be the set of all families of open subsets of X satisfying

1) if $U \in \mathcal{U} \in \Omega(X)$ and V is an open subset of X containing U, then $V \in \mathcal{U}$;

2) if $\mathcal{U} \in \Omega(X)$ and $\mathcal{V}$ is a family of open subsets of X with $\cup\mathcal{V} \in \mathcal{U}$, then there exists a finite subfamily $\mathcal{V}'$ of $\mathcal{V}$ such that $\cup\mathcal{V}' \in \mathcal{U}$.

For each $\mathcal{U} \in \Omega(X)$ and open subset V of R, let $[\![\mathcal{U},V]\!] = \{f \in C(X) : f^{-1}(V) \in \mathcal{U}\}$. The topology on C(X) generated by the sets of the form $[\![\mathcal{U},V]\!]$ is called the Isbell topology on C(X), and is denoted by $C_{is}(X)$.

(a) The Isbell topology is a splitting topology, and is therefore a hypo-Ascoli topology.

(b) The space $C_k(X) \le C_{is}(X)$.

(c) (Papadopoulos [1986]) If F is an evenly continuous subset of C(X), then the closure of F in $C_{is}(X)$ is the same as the closure of F in $C_p(X)$, so that the Isbell topology on F is the same as the topology of pointwise convergence on F.

Chapter IV

CARDINAL FUNCTIONS

Throughout this chapter, except where indicated in the exercises, the range space is always the space $\mathbb{R}$ of real numbers with the usual metric ρ. Also α is always a hereditarily closed, compact network on the domain space X. There is no loss of generality in assuming that α is closed under finite unions.

The notation $|S|$ is used to denote the cardinality of a set S. Some standard cardinal functions include the following. The weight of a space X is defined by $w(X) = \omega + \min\{|\mathcal{B}|: \mathcal{B}$ is a base for X.$\}$. The density of X is given by $d(X) = \omega + \min\{|D|: D$ is a dense subset of $X\}$. Finally, the character of X is $\chi(X) = \sup\{\chi(X,x): x \in X\}$, where $\chi(X,x) = \omega + \min\{|\mathcal{B}_x|: \mathcal{B}_x$ is a base at $x\}$. These cardinal functions characterize well-known topological properties: X is first countable if and only if $\chi(X) = \omega$; X is second countable if and only if $w(X) = \omega$; and X is separable if and only if $d(X) = \omega$. Other cardinal functions are defined as they are used.

1. Netweight.

Recall that an α-network for X is a family $\mathcal{B}$ of subsets of X such that if $A \in \alpha$ and U is open in X with $A \subset U$, then there exists a $B \in \mathcal{B}$ with $A \subset B \subset U$. The netweight of X is defined by $nw(X) = \omega + \min\{|\mathcal{B}|: \mathcal{B}$ is a network for $X\}$, and the α-netweight of X is defined by $\alpha nw(X) = \omega + \min\{|\mathcal{B}|: \mathcal{B}$ is an α-network for $X\}$. In the case that α consists of all compact subsets of X, then the α-netweight of X is also called the k-netweight and an α-network on X is called a k-network. If the k-netweight of a space is equal to ω then the space is called an $\aleph_0$-space, and if the netweight of a space is ω then the space is sometimes called a cosmic space.

Theorem 4.1.1. For each X and each α, $nw(C_\alpha(X)) \le w(X)$.

Proof. Let $\mathcal{B}$ be a base for X such that $|\mathcal{B}| = w(X)$, and let $\mathcal{V}$ be a countable base for $\mathbb{R}$. The family $\{[B,V]: B \in \mathcal{B}$ and $V \in \mathcal{V}\}$ is a subbase for a topology τ on C(X). To see that $C_\alpha(X) \le C_\tau(X)$, let $A \in \alpha$ and $V \in \mathcal{V}$. It suffices to show that [A,V] is open in $C_\tau(X)$; so let $f \in [A,v]$. For each $a \in A$, there is a $B_a \in \mathcal{B}$ such that $a \in B_a \subset f^{-1}(V)$. Since A is compact, there exist $a_1,\dots,a_n \in A$ such that $A \subset B_{a_1} \cup \dots \cup B_{a_n}$. Now $[B_{a_1},V] \cap \dots \cap [B_{a_n},V]$ is an open subset of $C_\alpha(X)$ which contains f and is contained in [A,V]. Therefore $nw(C_\alpha(X)) \le nw(C_\tau(X)) \le w(C_\tau(X)) \le |\mathcal{B}| \cdot |\mathcal{V}| = w(X)$. ∎

Theorem 4.1.2. For each X and each α, $nw(C_\alpha(X)) = \alpha nw(X)$.

Proof. Let $\mathcal{B}$ be an α-network for X which is closed under finite intersections such that $|\mathcal{B}| = \alpha nw(X)$, and let $\mathcal{V}$ be a countable base for $\mathbb{R}$. Now it is straightforward to check that the family of all finite intersections of members of $\{[B,V]: B \in \mathcal{B}$ and $V \in \mathcal{V}\}$ is a network for $C_\alpha(X)$. Thus $nw(C_\alpha(X)) \le \alpha nw(X)$.

For the reverse inequality, let $\mathcal{F}$ be a network for $C_\alpha(X)$ such that $|\mathcal{F}| = nw(C_\alpha(X))$. For each $F \in \mathcal{F}$, define $F^* = \{x \in X: f(x) > 0$ for each $f \in F\}$, and let $\mathcal{F}^* = \{F^*: F \in \mathcal{F}\}$. To show that $\mathcal{F}^*$ is an α-network for X, let $A \in \alpha$ and let U be an open subset of X containing A. Take $f \in C(X)$ such that $f(A) = \{1\}$ and $f(X \setminus U) = \{0\}$. Then $W = [A,(0,\infty)]$ is a neighborhood of f in $C_\alpha(X)$. Therefore there is an $F \in \mathcal{F}$ such that $f \in F \subset W$. It is clear that $A \subset F^*$, so that it remains to show that $F^* \subset U$. Suppose on the contrary that there is an $x \in F^* \setminus U$. Since $x \notin U$, then $f(x) = 0$. But since $x \in F^*$ and $f \in F$, then $f(x) > 0$. This contradiction establishes that $F^* \subset U$. Therefore $\alpha nw(X) \le nw(C_\alpha(X))$. ∎

Corollary 4.1.3. Let X be any space.

(a) $C_k(X)$ has a countable network if and only if X has a countable k-network.

(b) $C_p(X)$ has a countable network if and only if X has a countable network.

It is also true that $C_k(X)$ has a countable k-newtork if and only if X has a countable k-network, and $C_p(X)$ has a countable k-network if and only if X is countable. The netweight of $C_\rho(X)$ is discussed in the next section.

2. Density and Cellularity. The cellularity of a space X is defined by $c(X) = \omega + \sup\{|\mathcal{U}|: \mathcal{U}$ is a pairwise disjoint family of nonempty open subsets of $X\}$. Also define the weak weight of X by $ww(X) = \omega + \min\{w(Y)$: there is a continuous bijection from X onto $Y\}$. Finally, let the α-weight of X be given by $w_\alpha(X) = \sup\{w(A): A \in \alpha\}$.

Theorem 4.2.1. For each X and each α, $w_\alpha(X) \leq c(C_\alpha(X)) \leq d(C_\alpha(X)) = ww(X)$.

Proof. To show that $d(C_\alpha(X)) = ww(X)$, since $C_p(X) \leq C_\alpha(X) \leq C_k(X)$, it suffices to show that $d(C_k(X)) \leq ww(X) \leq d(C_p(X))$. So let $\phi: X \to Y$ be a continuous bijection, where $w(Y) = ww(X)$. Then the induced function $\phi^*: C_k(Y) \to C_k(X)$ is almost onto, so that $d(C_k(X)) \leq d(C_k(Y)) \leq nw(C_k(Y)) \leq w(Y) = ww(X)$. To see that $ww(X) \leq d(C_p(X))$, let D be an infinite dense subset of $C_p(X)$ having cardinality $d(C_p(X))$. Define $\psi: X \to \mathbb{R}^D$ by $\pi_f \circ \psi(x) = f(x)$ for each $x \in X$ and $f \in D$. Now ψ is continuous, and is also one-to-one since D separates points of X. Therefore, $ww(X) \leq w(\psi(X)) \leq |D| = d(C_p(X))$.

Since $c(C_\alpha(X)) \leq d(C_\alpha(X))$, it remains to show that $w_\alpha(X) \leq c(C_\alpha(X))$; so let $A \in \alpha$. Since A is compact, $nw(A) = w(A)$. Also since α is hereditarily closed, then $C_\alpha(A)$ is metrizable by Theorem 1.2.5, so that $c(C_\alpha(A)) = nw(C_\alpha(A))$. Therefore, $w(A) = nw(A) \leq \alpha nw(A) = nw(C_\alpha(A)) = c(C_\alpha(A)) \leq c(C_\alpha(X))$. Since this is true for each $A \in \alpha$, then $w_\alpha(X) \leq c(C_\alpha(X))$. ∎

Corollary 4.2.2. The following are equivalent.

(a) $C_k(X)$ is separable.

(b) $C_p(X)$ is separable.

(c) X has a coarser separable metrizable topology.

A space has the countable chain condition provided its cellularity is ω. Since $C_p(X)$ is a dense subspace of $\mathbb{R}^X$, and $\mathbb{R}^X$ has the countable chain condition, then so does $C_p(X)$. On the other hand, $C_k(X)$ may not always have the countable chain condition. The next corollary illustrates this.

Corollary 4.2.3. If $C_k(X)$ has the countable chain condition, then each compact subspace of X is metrizable.

For the uniform topology on C(X), with respect to ρ, most of these cardinal functions are the same. This is because, for a metric space M, w(M) = nw(M) = knw(M) = d(M) = c(M). So only the density is considered in this case.

Theorem 4.2.4. For each X, $d(C_\rho(X)) = w(\beta X)$.

Proof. Let e: $X \to \beta X$ be the embedding of X into its Stone–Cech compactification. The the induced function $e^*: C_\rho(\beta X) \to C_\rho(X)$ is a dense embedding. Therefore $w(\beta X) = ww(\beta X) = d(C_k(\beta X)) = d(C_\rho(\beta X)) = d(C_\rho(X))$. ■

Since βX is only metrizable if X is already a compact metrizable space, then the following is true.

Corollary 4.2.5. For each X, $C_\rho(X)$ is separable if and only if X is a compact metrizable space.

The proof of Theorem 4.2.4 also shows that $d(C_\mu(X)) = c(C_\mu(X)) = w(\beta X)$ for any compatible uniformity μ on $\mathbb{R}$, even if it is not metrizable.

There is a property on topological groups which is related to cellularity. Let G be a topological group (under addition), and let m be an infinite cardinal number. Then G is <u>totally m-bounded</u> provided that for each neighborhood of U of the identity in G, there exists a subset S of G such that $|S| \leq m$ and $G = \{s+u: s \in S \text{ and } u \in U\}$. Now G can be characterized as being totally m-bounded if and only if it is isomorphic to a subgroup of a group with cellularity less than or equal to m. Since $C_p(X)$ has the countable chain condition, then it is always totally $\aleph_0$-bounded.

<u>Theorem 4.2.6.</u> The space $C_\alpha(X)$ is totally m-bounded if and only if $w_\alpha(X) \leq m$.

Proof. Suppose first that $C_\alpha(X)$ is totally m-bounded. If $A \in \alpha$ and if $i: A \to X$ is the inclusion map, then the induced function $i^*: C_\alpha(X) \to C_\alpha(A)$ is a group homomorphism which is a continuous surjection. Therefore, $C_\alpha(A)$ is a metrizable totally m-bounded group. It follows that $w(A) = d(C_\alpha(A)) \leq m$, so that $w_\alpha(X) \leq m$.

For the converse, assume that $w_\alpha(X) \leq m$ and let $Y = \Sigma\{A: A \in \alpha\}$ be the topological sum. If $p: Y \to X$ is the natural map, then the induced function $p^*: C_\alpha(X) \to C_\alpha(Y)$ is an embedding. Therefore it suffices to show that $c(C_\alpha(Y)) \leq m$. First note that $C_\alpha(Y)$ is homeomorphic to $\Pi\{C_\alpha(A): A \in \alpha\}$. Now since $d(C_\alpha(A)) = w(A) \leq m$ for each $A \in \alpha$, then $c(\Pi\{C_\alpha(A): A \in \alpha\}) \leq m$. ∎

<u>Corollary 4.2.7.</u> The following are equivalent.

(a) $C_k(X)$ is totally $\aleph_0$-bounded.

(b) Each compact subspace of X is metrizable.

(c) X is a compact-covering image of a metrizable space.

3. Pseudocharacter. The pseudocharacter of X is defined by $\psi(X) = \sup\{\psi(X,x): x \in X\}$, where $\psi(X,x) = w + \min\{|\mathcal{G}|: \mathcal{G}$ is a family of open sets of X with $\cap\mathcal{G}=\{x\}\}$. Also the diagonal degree of X is $\Delta(X) = \omega + \min\{|\mathcal{G}|: \mathcal{G}$ is a family of open sets in XxX with $\cap\mathcal{G}$ equal to the diagonal in XxX$\}$. Finally, the weak α-covering number of X is defined to be $w\alpha c(X) = \omega + \min\{|\beta|: \beta \subset \alpha$ and $\cup\beta$ is dense in X$\}$.

Theorem 4.3.1. For each X and each α, $\psi(C_\alpha(X)) = \Delta(C_\alpha(X)) = w\alpha c(X)$.

Proof. Since the diagonal degree equals the pseudocharacter for any topological group, then it remains only to show that $\psi(C_\alpha(X)) = w\alpha c(X)$. Now if f_0 is the constant 0 function, then $\{f_0\} = \cap\{[A_s,V_s]: s \in S\}$ for some $|S| = \psi(C_\alpha(X))$, $A_s \in \alpha$, and V_s open in $\mathbb{R}$. Define $\beta = \{A_s: s \in S\}$, and assume that there exists some x in X which is not in $\overline{\cup\beta}$. Then choose an $f \in C(X)$ such that $f(x) = 1$ and $f(\overline{\cup\beta}) = \{0\}$. But each $[A_s,V_s]$ contains f, so that $f = f_0$; which is impossible since $f(x) = 1$. Therefore $\overline{\beta} = X$, and thus $w\alpha c(X) \le \psi(C_\alpha(X))$.

For the reverse inequality, let $\beta \subset \alpha$ such that $\overline{\cup\beta} = X$ and $|\beta| = w\alpha c(X)$. If $Y = \Sigma\{A: A \in \beta\}$ is the topological sum and if $p: Y \to X$ is the natural map, then p is almost onto, so that the induced function $p^*: C_\alpha(X) \to C_\alpha(Y)$ is a continuous injection. It follows that $\psi(C_\alpha(X)) \le \psi(C_\alpha(Y)) = \psi(\Pi\{C_\alpha(A): A \in \beta\}) \le |\beta|\cdot\omega = w\alpha c(X)$. ∎

Corollary 4.3.2. The following are equivalent.

(a) Each point of $C_k(X)$ is a G_δ-set.

(b) Each compact subset of $C_k(X)$ is a G_δ-set.

(c) $C_k(X)$ has a G_δ diagonal.

(d) $C_k(X)$ is submetrizable.

(e) X is almost σ-compact.

Corollary 4.3.3. The following are equivalent.

(a) Each point of $C_p(X)$ is a G_δ-set.

(b) Each compact subset of $C_p(x)$ is a G_δ-set.

(c) $C_p(X)$ has a G_δ diagonal.

(d) $C_p(X)$ is submetrizable.

(e) $C_p(X)$ has coarser separable metrizable topology.

(f) X is separable.

Observe the duality, for the topology of pointwise convergence, between (e) and (f) of Corollary 4.3.3 and (b) and (c) of Corollary 4.2.2. Likewise, the compact-open topology has an analogous duality between (d) and (e) of Corollary 4.3.2 and the dual statements which will be established in Theorem 5.5.3.

Corollary 4.3.3 can be used to establish a property of spaces having countable networks. First note that the continuous image of a separable metric space has a countable network (because the image of a base is a network).

__Theorem 4.3.4.__ If X has a countable network, then there exist separable metric spaces M_1 and M_2 and continuous bijections $\phi_1: M_1 \to X$ and $\phi_2: X \to M_2$.

Proof. To obtain M_1, just take X with the topology generated by the countable network as subbase. To obtain M_2, note first that Corollary 4.1.3 shows that $C_p(X)$ also has a countable network, and is therefore separable. Then by Corollary 4.3.3, $C_p(C_p(X))$ is submetrizable. From the comment before Theorem 2.3.5, the diagonal map $\Delta: X \to C_p(C_p(X))$ is an embedding, so that X is submetrizable. ∎

It follows from Theorem 4.3.4 that a space which has a countable network has both a finer metrizable topology and a coarser metrizable topology.

__4.__ __Character.__ An __α-cover__ of a space X is a family of subsets of X such that every member of α is contained in some member of this family. Then the __α-Arens__ __number__ of X is defined by $\alpha a(X) = \omega + \min\{|\mathcal{U}|: \mathcal{U} \subset \alpha$ and $\mathcal{U}$ is an α-cover of X$\}$. When α is the set of compact subsets of X, then an α-cover is called a k-cover, and X is said to be hemicompact whenever $\alpha a(X) = \omega$. Also when α is the set of finite subsets of X, then an α-cover is called an ω-cover, and in this case $\alpha a(X) = |X|$.

If $x \in X$, a collection $\mathcal{V}$ of nonempty open subsets of X is called a local π-base at x provided that for each open neighborhood U of x, there exists a $V \in \mathcal{V}$ which is contained in U. Define the π-character of X by $\pi\chi(X) = \omega + \sup\{\pi\chi(X,x): x \in X\}$, where $\pi\chi(X,x) = \min\{|\mathcal{V}|: \mathcal{V}$ is a local π-base at x$\}$.

The next theorem gives a characterization of the character of $C_\alpha(X)$. Since for any topological group, the character and the π-character are the same, then this theorem also characterizes the π-character of $C_\alpha(X)$.

Theorem 4.4.1. For each X and each α, $\chi(C_\alpha(X)) = \pi\chi(C_\alpha(X)) = \alpha a(X)$.

Proof. Let $\{[A_t,V_t]: t \in T\}$ be the local base at the zero function f_0 such that $|T| \leq \chi(C_\alpha(X))$. Suppose, by way of contradiction, that there exists an $A \in \alpha$ such that A is not contained in A_t for all $t \in T$. Then there is a $t \in T$ so that $[A_t,V_t] \subset [A,(-1,1)]$. Let $a \in A \setminus A_t$ and choose an $f \in C(X)$ such that $f(a) = 1$ and $fA_t) = \{0\}$. But then $f \in [A_t,V_t] \setminus [A,(-1,1)]$. With this contradiction, then it is seen that $A \subset A_t$ for some $t \in T$. Therefore $\alpha a(x) \leq |T| \leq \chi(C_\alpha(X))$.

For the reverse inequality, let $\{A_t: t \in T\} \subset \alpha$ be an α-cover of X such that $|T| \leq \alpha a(X)$. Let Y be the topological sum of the A_t's and let $p: Y \rightarrow X$ be the natural projection. Then $p^*: C_\alpha(X) \rightarrow C_\alpha(Y)$ is an embedding, so that $\chi(C_\alpha(X)) \leq \chi(C_\alpha(Y)) = \chi(\Pi\{C_\alpha(A_t): t \in T\}) \leq |T| \leq \alpha a(X)$. ∎

The next result includes the countable version of Theorem 4.4.1. This can be expressed in terms of the concept of a q-space, which means a space X for which each point has a sequence $\{U_n: n \in \omega\}$ of neighborhoods such that whenever $x_n \in U_n$ for each n, then the set $\{x_n: n \in \omega\}$ clusters in X.

<u>Theorem</u> <u>4.4.2.</u> The following are equivalent.

(a) $C_\alpha(X)$ is a q-space.

(b) $C_\alpha(X)$ is first countable.

(c) $C_\alpha(X)$ is metrizable.

(d) $\alpha a(X) = \omega$.

Proof. To prove that (a) implies (d), let $C_\alpha(X)$ be a q-space. Then the zero function f_0 has a sequence $\{B_n: n \in \omega\}$ of basic neighborhoods $B_n = [A_n, V_n]$ of f_0 which satisfies the definition of q-space at f_0. Suppose there exists an $x \in X \setminus \cup\{A_n: n \in \omega\}$. Then for each n, there is a $g_n \in C(X)$ such that $g_n(A_n) = \{0\}$ and $g_n(x) = n$. But then each $g_n \in B_n$ while $\{g_n: n \in \omega\}$ does not cluster in $C_\alpha(X)$. With this contradiction, it follows that $X = \cup\{A_n: n \in \omega\}$, so that $\psi(C_\alpha(X)) = \omega$ by Theorem 4.3.1. Because of this, there is no loss of generality in assuming that $\cap\{B_n: n \in \omega\} = \{f_0\}$ and the closure of each B_{n+1} is contained in B_n.

To show that $\{A_n: n \in \omega\}$ is an α-cover of X, suppose on the contrary that there is an $A \in \alpha$ such that A is not contained in A_n for each n. Then let $x_n \in A \setminus A_n$ for each n and choose $g_n \in C(X)$ such that $g_n(A_n) = \{0\}$ and $g_n(x_n) = 1$. Now each $g_n \in B_n \setminus [A,(-1,1)]$. But this means that f_0 is not a cluster point of $\{g_n: n \in \omega\}$. With this contradiction it follows that $\{A_n: n \in \omega\}$ is an α-cover of X and that $\alpha a(X) = \omega$ as desired.

It remains to prove that (d) implies (c). Let $\{A_n: n \in \omega\} \subset \alpha$ be an α-cover of X. To see that $C_\alpha(X)$ is metrizable, it suffices to embed it in a metrizable space. To this end, let Y be the topological sum of the A_n's and let $p: Y \to X$ be the natural projection. Then $p^*: C_\alpha(X) \to C_\alpha(Y)$ is an embedding. Since each $C_\alpha(A_n)$ is metrizable and $C_\alpha(Y)$ is homeomorphic to $\Pi\{C_\alpha(A_n): n \in \omega\}$, then $C_\alpha(Y)$ is metrizable. ■

It follows from Theorem 4.4.2 that $C_p(X)$ is metrizable if and only if X is countable and that $C_k(X)$ is metrizable if and only if X is hemicompact.

5. Weight. A collection $\mathcal{B}$ of nonempty open subsets of a space X is called a π-base for X provided that every nonempty open subset of X contains some member of $\mathcal{B}$. Then the π-weight of X is defined to be $\pi w(X) = \omega + \min\{|\mathcal{B}|: \mathcal{B}$ is a π-base for X$\}$. Also the α-α-netweight of X is given by $\alpha\alpha nw(X) = \omega + \min\{|\mathcal{B}|: \mathcal{B} \subset \alpha$ and $\mathcal{B}$ is an α-network on X$\}$.

Theorem 4.5.1. For each X and each α, $\alpha\alpha nw(X) = \alpha a(X) \cdot \alpha nw(X)$.

Proof. It follows from definitions that $\alpha nw(X) \leq \alpha\alpha nw(X)$. Also since every α-network is an α-cover, then $\alpha a(X) \leq \alpha\alpha nw(X)$. Therefore $\alpha a(X) \cdot \alpha nw(X) \leq \alpha\alpha nw(X)$.

For the reverse inequality, let $\tau = \alpha a(X) \cdot \alpha nw(X)$. Let $\mathcal{U} \subset \alpha$ be an α-cover of X with $|\mathcal{U}| \leq \tau$, and let $\mathcal{B}$ be a closed α-network of X with $|\mathcal{B}| \leq \tau$. Define $\mathcal{U} \cap \mathcal{B} = \{U \cap B: U \in \mathcal{U}$ and $B \in \mathcal{B}\}$, which is a subset of α. To see that $\mathcal{U} \cap \mathcal{B}$ is an α-network of X, let $A \in \alpha$ and let V be a neighborhood of A. Since $\mathcal{U}$ is an α-cover of X, then $A \subset U$ for some $U \in \mathcal{U}$. Also since $\mathcal{B}$ is an α-network of X, then $A \subset B \subset V$ for some $B \in \mathcal{B}$. Therefore $A \subset U \cap B \subset V$ and $U \cap B \in \mathcal{U} \cap \mathcal{B}$, so that $\mathcal{U} \cap \mathcal{B}$ is an α-network of X contained in α. It follows that $\alpha\alpha nw(X) \leq |\mathcal{U} \cap \mathcal{B}| \leq \tau$. ■

Obviously, if α consists of the finite subsets of X then $\alpha\alpha nw(X) = \omega$ if and only if X is countable. On the other hand, Theorem 4.5.1 says that whenever α consists of the compact subsets of X then $\alpha\alpha nw(X) = \omega$ if and only if X is a hemicompact $\aleph_0$-space.

Since the weight and the π-weight are identical for any topological group, the following theorem is proved only for the weight.

Theorem 4.5.2. For each X and each α, $w(C_\alpha(X)) = \pi w(C_\alpha(X)) = \alpha\alpha nw(X)$.

Proof. For any topological space, the weight is equal to the character times the netweight. Therefore $w(C_\alpha(X)) = \chi(C_\alpha(X)) \cdot nw(C_\alpha(X))$. Since $\chi(C_\alpha(X)) = \alpha a(X)$ by Theorem 4.4.1 and since $nw(C_\alpha(X)) = \alpha nw(X)$ by Theorem 4.1.2, then the result follows from Theorem 4.5.1. ■

Corollary 4.5.3. For each X, $C_k(X)$ is second countable if and only if X is a hemicompact $\aleph_0$-space.

Corollary 4.5.4. For each X, $C_p(X)$ is second countable if and only if X is countable.

6. Weak Weight. Recall that the weak weight of a space X is defined in section 2, and is used to characterize the density of $C_\alpha(X)$. In this section, the weak weight of $C_\alpha(X)$ is characterized. In order to facilitate this characterization, the notion of the logarithm of an infinite cardinal number is used. In particular, $\log(\alpha) = \min\{\beta\colon \alpha \leq 2^\beta\}$.

Theorem 4.6.1. For each X and each α, $ww(C_\alpha(X)) = w\alpha c(X) \cdot \log(\alpha nw(X))$.

Proof. First observe that $w\alpha c(X) = \psi(C_\alpha(X))$ by Theorem 4.3.1, and $\alpha nw(X) = nw(C_\alpha(X))$ by Theorem 4.1.2. Now for any space Z, $\psi(X) \cdot \log(nw(Z)) \leq ww(Z)$. To see this, note that $\psi(Z) \leq ww(Z)$ is immediate. To establish that $\log(nw(Z)) \leq ww(Z)$, let $\phi\colon Z \to Y$ be a continuous bijection such that $w(Y) = ww(Z)$. Then $nw(Z) \leq |Z| = |Y| \leq 2^{w(Y)} = 2^{ww(Z)}$, so that $\log(nw(Z)) \leq ww(Z)$.

On the other hand, to show that $ww(C_\alpha(X)) \leq w\alpha c(X) \cdot \log(\alpha nw(X))$, let τ =

$w\alpha c(X)$ and $\sigma = \log(\alpha nw(X))$. So let $\{A_t: t \in T\} \subset \alpha$ such that $|T| \le \tau$ and $\cup\{A_t: t \in T\}$ is dense in X. Since each $A_t \in \alpha$ then $C_\alpha(A_t)$ is metrizable, and in particular $w(C_\alpha(A_t)) = d(C_\alpha(A_t))$. Since the inclusion map $i: A_t \to X$, for each A_t, induces a continuous surjection $i^*: C_\alpha(X) \to C_\alpha(A_t)$, then $d(C_\alpha(A_t)) \le d(C_\alpha(X))$. But $d(C_\alpha(X)) \le nw(C_\alpha(X)) = \alpha nw(X) \le 2^\sigma$. Now if a metric space has weight $\le 2^\sigma$, then it has weak weight $\le \sigma$. This is because it can be embedded in the ω power of the hedgehog space with 2^σ spines. This hedgehog space has a natural continuous bijection onto $2^\sigma \times [0,1]$, where 2^σ is the product of σ copies of the discrete two point space with the product topology, and thus it has weak weight $\le \sigma$. Therefore $ww(C_\alpha(A_t)) \le \sigma$ for each $t \in T$.

Let Y be the topological sum of the A_t's, and let $p: Y \to X$ be the natural map. Then $p^*: C_\alpha(X) \to C_\alpha(Y)$ is a continuous injection, so that $ww(C_\alpha(X)) \le ww(C_\alpha(Y)) = ww(\Pi\{C_\alpha(A_t): t \in T\}) \le \Sigma\{ww(C_\alpha(A_t)): t \in T\} \le \tau\sigma$. ∎

Another way of saying Theorem 4.6.1 is that $ww(C_\alpha(X)) \le \tau$ if and only if $w\alpha c(X) \le \tau$ and $\alpha nw(X) \le 2^\tau$.

<u>Corollary 4.6.2.</u> For each X, $ww(C_p(X)) = d(X)$. In particular, $C_p(X)$ has a coarser separable metrizable topology if and only if X is separable.

<u>Corollary 4.6.3.</u> For each X, $C_k(X)$ has a coarser separable metrizable topology if and only if X is almost σ-compact and $knw(X) \le 2^\omega$.

<u>Corollary 4.6.4.</u> For each X, $ww(C_\rho(X)) = \log(w(\beta X))$.

<u>7. Tightness and the Fréchet Property.</u> The <u>tightness</u> of a space X is defined by $t(X) = \sup\{t(X,x): x \in X\}$, where $t(X,x) = \omega + \min\{\tau$: for each $Y \subset X$ with $x \in \overline{Y}$ there is $A \subset Y$ with $|A| \le \tau$ and $x \in \overline{A}\}$. Then X has countable tightness if and only if $t(X) = \omega$.

Also define the <u>α-Lindelöf degree</u> of X by $\alpha L(X) = \omega + \min\{\tau$: every open α-cover of X has an α-subcover of cardinality $\le \tau\}$. If α consists of the singleton subsets of X, then the α-Lindelöf degree of X is called the <u>Lindelöf degree</u> of X and written L(X).

<u>Theorem 4.7.1.</u> For each X and each α, $t(C_\alpha(X)) = \alpha L(X)$.

Proof. Let $\tau = t(C_\alpha(X))$, and let $\mathcal{U}$ be any open α-cover of X. Then for each $A \in \alpha$, there is a $U_A \in \mathcal{U}$ such that $A \subset U_A$. Now for each such A, choose an $f_A \in C(X)$ such that $f_A(A) = \{0\}$ and $f_A(X \setminus U_A) = \{1\}$. If $F = \{f_A\colon A \in \alpha\}$, then the zero function, f_0, is in the closure of F in $C_\alpha(X)$. So there is an F' contained in F, with cardinality $\le \tau$, which contains f_0 in its closure. Define $\mathcal{V} = \{U_A\colon f_A \in F'\}$. To see that $\mathcal{V}$ is an α-subcover of $\mathcal{U}$, let $A \in \alpha$. Then define $W = [A,(-1,1)]$, which is a neighborhood of f_0. So for some $B \in \alpha$, $f_B \in F' \cap W$. Now for any $x \in A$, $f_B(x) < 1$; while for any $x \in X \setminus U_B$, $f_B(x) = 1$. Therefore $A \subset U_B$, so that $\mathcal{V}$ is an α-subcover of $\mathcal{U}$ having cardinality $\le \tau$. This establishes that $\alpha L(X) \le t(C_\alpha(X))$.

To show the reverse inequality, let $\sigma = \alpha L(X)$. Since $C_\alpha(X))$ is homogeneous, it suffices to show that $t(C_\alpha(X),f_0) \le \sigma$. To this end, let G be a subset of $C_\alpha(X)$ containing f_0 in its closure. For each $n \in \omega$ and each $A \in \alpha$, choose a $g_{n,A} \in G \cap [A,(-1/n,1/n)]$, and set $W(n,A) = \{x \in X\colon |g_{n,a}(x)| < 1/n\}$. Then for each n, the family $\mathcal{W}_n = \{W(n,A)\colon A \in \alpha\}$ is an open α-cover of X. So each $\mathcal{W}_n$ has an α-subcover $\mathcal{V}_n$ of cardinality $\le \sigma$. Define $G' = \{f_{n,A}\colon n \in \omega$ and $W(n,A) \in \mathcal{V}_n\}$. It is evident that $G' \subset G$, $|G'| \le \sigma$, and f_0 is the closure of G' in $C_\alpha(X)$. Therefore $t(C_\alpha(X)) \le \sigma$ as desired. ■

<u>Corollary 4.7.2.</u> The space $C_k(X)$ has countable tightness if and only if every open k-cover of X has a countable k-subcover.

Corollary 4.7.3. For each X, $t(C_p(X)) = \sup\{L(X^n): n \in \omega\}$. In particular, $C_p(X)$ has countable tightness if and only if X^n is Lindelöf for each $n \in \omega$.

Proof. Suppose first that $L(X^n) \leq \tau$ for each $n \leq \omega$. Let $\mathcal{U}$ be an open ω-cover of X. Then for each n, define $\mathcal{U}_n = \{U^n: U \in \mathcal{U}\}$, which is an open cover of X^n. Now let $\mathcal{U}_n'$ be a subcover of $\mathcal{U}_n$ of cardinality $\leq \tau$. Then the family {U: for some $n \in \omega$, $U^n \in \mathcal{U}_n'$} is an ω-subcover of $\mathcal{U}$ having cardinality $\leq \tau$.

For the converse, suppose that each open ω-cover of X has an ω-subcover with cardinality $\leq \tau$. If $\mathcal{W}$ is an open cover of X^n, then let $\mathcal{V} = \{V \subset X$: V is open and V^n is covered by finitely many members of $\mathcal{W}\}$. Since $\mathcal{V}$ is an open ω-cover of X, it has an ω-subcover $\mathcal{V}'$ having cardinality $\leq \tau$. Then the family $\{V^n: V \in \mathcal{V}'\}$ is an open cover of X^n with cardinality $\leq \tau$. It follows that $L(X^n) \leq \tau$. ∎

A property which is stronger than having countable tightness but weaker than being first countable is that of being a Fréchet space. A space X is a Fréchet space provided that if $x \in X$ and if A is a subset of X containing x in its closure, then there exists a sequence $\{x_n: n \in \omega\}$ in A which converges to x in X.

In order to characterize $C_\alpha(X)$ being a Fréchet space the following concept is needed. An α-sequence in X is any sequence $\{C_n: n \in \omega\}$ of subsets of X with the property that if $A \in \alpha$ then there exists an $m \in \omega$ such that $A \subset C_n$ for all $n \geq m$.

Theorem 4.7.4. The space $C_\alpha(X)$ is a Fréchet space if and only if every open α-cover of X contains an α-sequence.

Proof. Suppose first that $C_\alpha(X)$ is a Fréchet space, and let $\mathcal{U}$ be an open α-cover of X. For each $A \in \alpha$ there is a $U_A \in \mathcal{U}$ such that $A \subset U_A$; and there is an $f_A \in C(X)$ such that $f_A(A) = \{0\}$ and $f_A(X \setminus U_A) = \{1\}$. It is straightforward to check that the zero function, f_0, is in the closure of $\{f_A: A \in \omega\}$. Then there is a sequence

$\{f_{A_n}: n \in \omega\}$ which converges to f_0, so that $\{U_{A_n}: n \in \omega\}$ is an α-sequence contained in $\mathcal{U}$.

For the converse, let G be a subset of $C_\alpha(X)$ containing f_0 in its closure, and suppose that every open α-cover contains an α-sequence. For each $n \in \omega$ and $A \in \alpha$, define $f_{n,A}$, $W(n,A)$, and $\mathcal{W}_n$ as in the proof of Theorem 4.7.1. In particular, each $\mathcal{W}_n$ is an open α-cover of X. Now define a sequence $\{\mathcal{U}_n: n \in \omega\}$ of open α-covers of X as follows. Let $\mathcal{U}_1 = \mathcal{W}_1$, and for each $n > 1$, let $\mathcal{U}_n$ be an open α-cover which refines both $\mathcal{U}_{n-1}$ and $\mathcal{W}_n$. It may be assumed that $X \notin \alpha$. Observe that the family $\{X \setminus \{x\}: x \in X\}$ is an open α-cover of X, so by hypothesis there is a sequence $\{x_n: n \in \omega\}$ in X such that $\{X \setminus \{x_n\}: n \in \omega\}$ is an α-sequence. Next, define $\mathcal{U}_n' = \{U \setminus \{x_n\}: V \in \mathcal{U}_n\}$ for each n, and let $\mathcal{V} = \cup\{\mathcal{U}_n': n \in \omega\}$, which is an open α-cover of X. Now choose an α-sequence $\{V_n: n \in \omega\}$ from $\mathcal{V}$.

For each $k \in \omega$ there is an $n_k \in \omega$ such that $V_k \subset U_k$ for some $U_k \in \mathcal{U}_{n_k}$. Therefore, for some $A_k \in \alpha$, $V_k \subset W(n,A_k)$. But, if $n \in \omega$ and $\cup\{A_i: i \leq n\} \subset V_k$, then $n_k > n$. Hence the sequence $\{n_k: k \in \omega\}$ is an infinite sequence, and it follows that $\{U_k: k \in \omega\}$ is an α-sequence in X with $U_k \in \mathcal{U}_{n_k}$. Take an increasing subsequence $\{n_{k_i}: i \in \omega\}$, and let $f_i = f_{n_{k_i}, A_{k_i}}$. Then it is straightforward to check that the sequence $\{f_i: i \in \omega\}$ converges to f_0. ∎

<u>8. Hereditary Density and Hereditary Lindelöf Degree.</u> The <u>hereditary density</u> of X is defined by $hd(X) = \sup\{d(A): A \subset X\}$, and the <u>hereditary Lindelöf degree</u> of X is $hL(X) = \sup\{L(A): A \subset X\}$. In order to get estimates of $hd(C_\alpha(X))$ and $hL(C_\alpha(X))$, it will be necessary to consider α itself as a topological space. The topology used on α is the Vietoris topology. Subbasic open sets in α are either $\langle V \rangle$ or $\langle V,X \rangle$, where V is open in X and $\langle V \rangle = \{A \in \alpha: A \text{ is contained in } V\}$ and $\langle V,X \rangle = \{A \in \alpha: A \text{ intersects } V\}$.

A special theorem is needed for this setting. If $f: X \times Y \to Z$ is a function, then f is <u>continuous on</u> X provided that for each $y \in Y$, the restriction $f_y: X \to Z$ of f to X is continuous, where f_y is defined by $f_y(x) = f(x,y)$ for each $x \in X$. A similar definition defines f being continuous on Y.

Theorem 4.8.1. If $f\colon X \times Y \to Z$ is a function which is continuous on Y and such that X has the weakest topology making f continuous on X, then

$$\text{(a)} \qquad hd(X) \le hL(Y) \cdot w(Z),$$

$$\text{(b)} \qquad hL(X) \le hd(Y) \cdot w(Z).$$

Proof. Let $\mathcal{B}$ be a base for Z having minimal cardinality. For each $B \in \mathcal{B}$ and $x \in X$, define $B^x = \{y \in Y\colon f(x,y) \in B\}$; and for each $B \in \mathcal{B}$ and $y \in Y$, define $B_y = \{x \in X\colon f(x,y) \in B\}$.

(a) Let $A \subset X$. Since f is continuous on Y, for each $B \in \mathcal{B}$, $\{B^a\colon a \in A\}$ is a family of open subsets of Y. So for each $B \in \mathcal{B}$, there exists $A_B \subset A$ such that $|A_B| \le hL(Y)$ and $\cup\{B^a\colon a \in A_B\} = \cup\{B^a\colon a \in A\}$. Define $A' = \cup\{A_B\colon B \in \mathcal{B}\}$, so that $|A'| \le hL(Y) \cdot w(Z)$. To show that A' is dense in A, let $a \in A$ and let U be a neighborhood of a in X. Since X has the weakest topology making f continuous on X, there exists a $B \in \mathcal{B}$ and $y \in Y$ such that $a \in B_y \subset U$. But then $y \in B^a$, so that there exists a $b \in A_B$ with $y \in B^b$. Therefore $b \in B_y$, and hence $b \in A_B \cap U \subset A' \cap U$.

(b) Since X has the weakest topology making f continuous on X, then the family $\mathcal{A} = \{B_y\colon B \in \mathcal{B} \text{ and } y \in Y\}$ is a base for X. So let $\mathcal{A}' \subset \mathcal{A}$. For each $B \in \mathcal{B}$, define $C_B = \{y \in Y\colon B_y \in \mathcal{A}'\}$. Now for each such B, there exists a dense subset D_B of C_B such that $|D_B| \le hd(Y)$. Then define $\mathcal{A}'' = \{B_y\colon B \in \mathcal{B} \text{ and } y \in D_B\}$. Since $\mathcal{A}'' \subset \mathcal{A}'$, then $\cup\mathcal{A}'' \subset \cup\mathcal{A}'$. To show containment in the other direction, let $x \in \cup\mathcal{A}'$. Then there exist $B \in \mathcal{B}$ and $y \in Y$ such that $x \in B_y \in \mathcal{A}'$. But then $y \in C_B \cap B^x$, so that there exists a $d \in D_B \cap B^x$. Therefore $x \in B_d \in \mathcal{A}''$, and thus $x \in \cup\mathcal{A}''$. ∎

Corollary 4.8.2. If Z is second countable and $f: X \times Y \to Z$ is a function such that the topology on each factor of $X \times Y$ is the weakest topology making f continuous on that factor, then $hd(X) = hL(Y)$ and $hL(X) = hd(Y)$.

Theorem 4.8.3. Let α be a compact network on X. If α is considered as a topological space with the Vietoris topology, then

$$(1) \qquad hd(X) \le hL(C_\alpha(X)) = hd(\alpha),$$

$$(2) \qquad hL(X) \le hd(C_\alpha(X)) = hL(\alpha).$$

Proof. Let $e: X \times C_\alpha(X) \to \mathbb{R}$ be the evaluation function. Recall that for each $x \in X$, $e_x: C_\alpha(X) \to \mathbb{R}$ is continuous. Also for each $f \in C_\alpha(X)$, let e_f: be the restriction of e to X. Since for each $f \in C_\alpha(X)$ and each open V in $\mathbb{R}$, $e_f^{-1}(V) = f^{-1}(V)$, and since X is completely regular, then the topology on X is the weakest topology making e continuous on X. Therefore by Theorem 4.8.1, $hd(X) \le hL(C_\alpha(X))$ and $hL(X) \le hd(C_\alpha(X))$.

Now let K be the space of compact subsets of $\mathbb{R}$ with the Vietoris topology. This space has a countable base. Define $\theta: \alpha \times C_\alpha(X) \to K$ by $\theta(A,f) = f(A)$ for each $A \in \alpha$ and $f \in C_\alpha(X)$. For each $A \in \alpha$ and $f \in C_\alpha(X)$, let $\theta_A: C_\alpha(X) \to K$ and $\theta_f: \alpha \to K$ be the restrictions of θ to $C_\alpha(X)$ and α. It is straightforward to show that for each $f \in C_\alpha(X)$ and each open V in $\mathbb{R}$, $\theta_f^{-1}(\langle V \rangle) = \langle f^{-1}(V) \rangle$ and $\theta_f^{-1}(\langle V,\mathbb{R} \rangle) = \langle f^{-1}(V),X \rangle$. This shows that θ is continuous on α. Each open $\langle U \rangle$ in α can be written as a union of sets of the form $\langle f^{-1}(V) \rangle$, and each open $\langle U,X \rangle$ in α can be written as a union of sets of the form $\langle f^{-1}(V),X \rangle$. So α has the weakest topology making θ continuous on α. Also it is straightforward to show that for each $A \in \alpha$ and V open in $\mathbb{R}$, $\theta_A^{-1}(\langle V \rangle) = [A,V]$. It remains to show that $\theta_A^{-1}(\langle V,\mathbb{R} \rangle)$ is open in $C_\alpha(X)$; then the topology on $C_\alpha(X)$ would be the weakest topology making θ continuous on $C_\alpha(X)$. If $f \in \theta_A^{-1}(\langle V,\mathbb{R} \rangle)$, then there exists an

$x \in A \cap f^{-1}(V)$. Therefore $f \in [x,V] \subset \theta_A^{-1}(<V,\mathbb{R}>)$. Now Corollary 4.8.2 applies to show that $hL(C_\alpha(X)) = hd(\alpha)$ and $hd(C_\alpha(X)) = hL(\alpha)$. ∎

Corollary 4.8.4. For any X, $hL(C_p(X)) = hd(X)$ and $hd(C_p(X)) = hL(X)$.

9. Exercises and Problems for Chapter IV.

1. Metrizability. Let X and R be spaces.

(a) $C_k(X,R)$ is metrizable if and only if X is hemicompact and R is metrizable.

(b) $C_p(X,R)$ is metrizable if and only if X is countable and R is metrizable.

2. Countable Chain Condition.

(a) (Vidossich [1972]) If X is submetrizable, then $C_k(X)$ has countable chain condition. The converse fails (take $X = \omega_1$).

(b) (Arhangelskii [1982]) For any space X, $c(X) \leq \sup\{w(F): F$ is a compact subspace of $C_p(X)\}$. In particular, if each compact subset of $C_p(X)$ is metrizable, then X has countable chain condition. For an infinite compact space X, $c(X) = \sup\{w(F): F$ is a compact subspace of $C_p(X)\}$.

3. Lindelöf Spaces. (McCoy [1980a], Arhangelskii [1982])

(a) For each space X, $\sup\{t(X^n): n \in \omega\} \leq L(C_p(X))$. In particular, if $C_p(X)$ is a Lindelöf space, then X^n has countable tightness for each $n \in \omega$.

(b) Every discrete family of open subsets of X has cardinality less than or equal to $L(C_p(X))$. In particular, if $C_p(X)$ is a Lindelöf space, then every discrete family of open subsets of X is countable.

(c) If X is normal and $C_p(X)$ is a Lindelöf space, then every discrete family of closed subsets of X is countable.

(d) The space $C_p(X)$ is Lindelöf if and only if $C_p(X)$ is paracompact.

4. Σ–products. (Gul'ko [1977],[1978])

(a) If X is a closed subspace of a Σ–product of separable metric spaces such that every compact subset is metrizable and if R is a separable metric space, then $C_k(X,R)$ is a Lindelöf space.

(b) If X is a closed subspace of a Σ–product of separable metric spaces and if R is a separable metric space, then $C_p(X,R)$ is a Lindelöf space.

(c) If X is a closed subspace of a Σ–product of separable metric spaces and if R is a metric space, then $C_p(X,R)$ is a paracompact space.

5. Normal Spaces.

(a) (Arhangelskii and Tkačuk [1985]) If $C_p(X)$ is normal, then every compact subspace of X has countable tightness.

(b) (Arhangelskii and Tkačuk [1985]) If $C_p(X)$ is normal, then $C_p(Y)$ is normal for every closed C–embedded subspace Y of X.

(c) (Pol [1974]) If X is a metric space, then $C_p(X,[0,1])$ is normal if and only if the subspace of X consisting of the non–isolated points is separable.

6. Paracompact Spaces. (O'Meara [1971]) If X has a countable k–network and R is paracompact and has a σ–locally finite k–network, then $C_k(X,R)$ is paracompact and has a σ–locally finite k–network.

7. General Range. Let X and R be spaces.

(a) $\alpha nw(X)\cdot nw(R) \le nw(C_\alpha(X,R) \le \alpha nw(X)\cdot w(R)$. So these are equalities if R is metrizable.

(b) $d(R) \le d(C_\alpha(X,R))$.

(c) (Noble [1974]) If R is a retract of a convex subset of some locally convex linear space, then $d(C_\alpha(X,R)) = ww(X)\cdot d(R)$.

(d) If R is second countable, then $hd(X) \le hL(C_\alpha(X,R)) \le hd(\alpha)$ and $hL(X) \le$

$hd(C_\alpha(X,R)) \leq hL(\alpha)$.

8. Fréchet Spaces and Tightness.

(a) If $C_p(X)$ is a Fréchet space, then for every sequence $\{\mathcal{U}_n: n \in \omega\}$ of open covers of X there is a $U_n \in \mathcal{U}_n$ for each n such that the collection $\{U_n: n \in \omega\}$ covers X. Deduce that $C_p([0,1])$ is not a Fréchet space.

(b) (Gerlits and Nagy [1982]) If $C_p(X)$ is a Fréchet space, then X is 0-dimensional.

(c) (Tkačuk [1984], Arhangelskii and Tkačuk [1985]) The space $C_p(X)$ is a Fréchet space if and only if $(C_p(X))^\omega$ is a Fréchet space.

(d) (Tkačuk 1986]) $t(C_p(X) = t((C_p(X))^\omega)$.

9. Fan Tightness and Hurewicz Spaces. Corollary 4.7.3 has an analog for a cardinal invariant called <u>fan tightness</u>. A space X has <u>countable fan tightness</u> provided that for each $x \in X$ and each sequence $\{A_n : n \in \omega\}$ of subsets of X with $x \in \cap\{\overline{A_n} : n \in \omega\}$, there exist finite subsets $B_n \subset A_n$, for each n, such that x is in the closure of $\cup\{B_n : n \in \omega\}$. If each B_n is replaced with a singleton subset in the above definition, then X is said to have <u>countable strong fan tightness</u>. A space X is called a <u>Hurewicz space</u> provided that for each sequence $\{\mathcal{U}_n : n \in \omega\}$ of open covers of X, there exist finite subfamilies $\mathcal{V}_n \subset \mathcal{U}_n$, for each n, such that $\cup\{\mathcal{V}_n : n \in \omega\} = X$. If each $\mathcal{U}_n$ is replaced with a singleton subfamily in the previous definition, then X is said to have <u>property</u> C". Clearly if X has property C" then X is a Hurewicz space, and if X is a Hurewicz space then X is a Lindelöf space.

(a) (Arhangelskii [1986]) For any space X, $C_p(X)$ has countable fan tightness if and only if X^n is a Hurewicz space for all $n \in \omega$.

(b) (Sakai [1987]) For any space X, $C_p(X)$ has countable strong fan tightness if and only if X^n has property C" for all $n \in \omega$.

10. Lindelöf Σ-spaces. (Arhangelskii [1982], Ntantu [1985])

(a) If X is almost σ-compact, then $C_k(X)$ is a Lindelöf Σ-space if and only if X has countable k-netweight.

(b) If $C_p(X)$ is a Lindelöf Σ-space and A is a compact subspace of X, then w(A) = c(A).

(c) If $C_p(X)$ is a Lindelöf Σ-space, then the closure of every countable subset of X has a countable network. Conclude that if X is separable, then $C_p(X)$ is a Lindelöf Σ-space if and only if X has a countable network.

(d) If X is compact and $C_p(X)$ is a Lindelöf Σ-space, then X contains a dense first countable subspace.

(e) If X is compact and $C_p(X)$ contains a dense σ-compact subspace, then $C_p(X)$ is a Lindelöf Σ-space.

11. k-spaces. (Pytkeev [1982], Gerlits [1983]) $C_\alpha(X)$ is a k-space if and only if $C_\alpha(X)$ is a Fréchet space.

12. σ-spaces and Σ-spaces. (Ntantu [1985]) A space is a <u>σ-space</u> if it has a σ-discrete network. A space X is a <u>(strong)</u> <u>Σ-space</u> provided there exists a σ-discrete family $\mathcal{F}$ and a cover $\mathcal{C}$ by closed countably compact (compact) sets such that whenever $C \in \mathcal{C}$ and $C \subset U$ with U open in X, then $C \subset F \subset U$ for some $F \in \mathcal{F}$. Every cosmic space is a σ-space, every σ-space is a strong Σ-space, and every strong Σ-space is a Σ-space.

(a) If X has a countable network (i.e., is cosmic) and R is metrizable, then $C_p(X,R)$ is a σ-space (extend the proof of Theorem 4.1.2).

(b) If $C_\alpha(X)$ is a hereditarily strong Σ-space, then $w\alpha c(X) = \aleph_0$. Deduce that $C_\alpha(X)$ is a σ-space if and only if $C_\alpha(X)$ is a hereditarily strong Σ-space.

(c) If X is a Σ-space, then $C_\alpha(X)$ is a hereditarily separable Σ-space if and only if $C_\alpha(X)$ is a cosmic space.

13. Point Countable Type. If $A \subset X$, define the <u>character of</u> A in X by $\chi(X,A) = \omega + \min\{|\mathcal{B}|: \mathcal{B}$ is a base at A$\}$; where $\mathcal{B}$ is a base at A provided the members of $\mathcal{B}$ are open sets containing A, and every neighborhood of A contains some member of $\mathcal{B}$. Also define space X to have <u>point countable type</u> provided each point of X is contained in a compact subset of X having countable character in X.

(a) If D is dense in X and $A \subset D$, then $\chi(D,A) = \chi(X,A)$.

(b) The space $C_\alpha(X)$ contains a dense subspace of point countable type if and only if $C_\alpha(X)$ is of point countable type. Conclude that $C_\alpha(X)$ has a dense subspace of point countable type if and only if $C_\alpha(X)$ is metrizable.

14. Cardinality.

(a) (Hodel [1984]) For any space X, $|C(X)| \leq 2^{d(X)}$. In particular, $|C(X)| = 2^\omega$ for any separable space X.

(b) (Comfort [1971]) For any space X, $|C(X)|^\omega = |C(X)|$.

(c) (Comfort and Hager [1970]) For a space X, define $wc(X) = \omega + \min\{\tau$: every open cover of X has a subfamily of cardinality less than or equal to τ whose union is dense in X$\}$. Then $|C(X)| \leq (w(X))^{wc(X)}$. In particular, if X is a first countable Hausdorff space which either is a Lindelöf space or has ccc, then $|C(X)| = 2^\omega$.

15. Realcompact Spaces and the Hewitt Number. (Arhangelskii [1983], Uspenskii [1983]) For each cardinal τ, a subset A of X is of <u>type</u> G_τ in X provided it can be written as the intersection of τ open subsets of X. A subset A of X is <u>τ-embedded</u> in X provided that for every $x \in X \setminus A$ there exists a subset B of $X \setminus A$ which contains x and is of type G_τ in X. For each space X, the <u>Hewitt number</u> q(X) is defined by $q(X) = \omega + \min\{\tau$: X is τ-embedded in $\beta X\}$. The space X is realcompact if and only if $q(X) = \aleph_0$. A function $f: X \to Y$ is <u>τ-continuous</u> (<u>strictly τ-continuous</u>, respectively) provided that for every subset A of X with $|A| \leq \tau$, $f|_A$ is continuous ($f|_A$ has

continuous extension to X, respectively). The functional tightness of X is defined by $t_0(X) = \omega + \min\{\tau\colon$ every τ-continuous $f\colon X \to \mathbb{R}$ is continuous$\}$, and the weak functional tightness of X is given by $t_m(X) = \omega + \min\{\tau\colon$ every strictly τ-continuous $f\colon X \to \mathbb{R}$ is continuous$\}$. Note that if X is normal, then $t_m(X) = t_0(X)$.

(a) For any space X, $q(C_p(X)) = t_m(X)$.

(b) For any space X, $t_0(C_p(X)) = t_m(C_p(X)) = q(X)$.

16. Monolithic and Stable Spaces. (Arhangelskii [1984]) For every space X, $d(X) \leq nw(X)$ and $ww(X) \leq nw(X)$. A space X is called monolithic provided that for each subspace A of X, $d(A) = nw(A)$; and X is called stable provided that for each continuous image Y of X, $ww(Y) = nw(Y)$. The space $C_p(X)$ is monolithic if and only if X is stable; and $C_p(X)$ is stable if and only if X is monolithic.

Chapter V

COMPLETENESS AND OTHER PROPERTIES

The first four sections of this chapter study the completeness properties of $C_\alpha(X) = C_\alpha(X,\mathbb{R})$, where α is a hereditarily closed compact network on X. The properties range from complete metrizability to the Baire space property. The remaining two sections deal with subsets of function spaces having certain properties.

1. Uniform Completeness.

A space X is called an α_R-space provided that every real-valued function on X is continuous whenever its restriction to each member of α is continuous. If α is the family of compact subsets of X, then an α_R-space is called a k_R-space; and if α is the set of finite subsets of X, then X is an α_R-space if and only if X is a discrete space.

Recall from Chapter I that for a hereditarily closed and compact network α, the topology of $C_\alpha(X)$ is generated by the uniformity of uniform convergence on α. One kind of completeness on $C_\alpha(X)$ is for this uniformity to be complete. In this case, $C_\alpha(X)$ is called uniformly complete. If $C_\alpha(X)$ is completely metrizable, then it is uniformly complete. Now completeness with respect to some some specified uniformity is not a topological property and would thus not normally be considered as a completeness property. However, uniform completeness of $C_\alpha(X)$ can be characterized in terms of a topological property on X with respect to α.

Theorem 5.1.1. The space $C_\alpha(X)$ is uniformly complete if and only if X is an α_R-space.

Proof. Suppose $C_\alpha(X)$ is uniformly complete, and let f be a real-valued function on X such that $f|_A$ is continuous for each $A \in \alpha$. Let $f_A \in C(X)$ be an extension of $f|_A$ for each A. Then the net $(f_A)_{A\in\alpha}$ is Cauchy in $C_\alpha(X)$ and therefore converges (to f) in $C_\alpha(X)$.

Conversely, let X be an α_R-space, and let (f_i) be a Cauchy net in $C_\alpha(X)$. If $A \in \alpha$, then the net $(f_i|_A)$ is Cauchy in $C_\alpha(A) = C_k(A)$. Since $C_k(A)$ is a complete metric space, then $(f_i|_A)$ converges to some f_A in $C_k(A)$. Define f: $X \to \mathbb{R}$ by $f(x) = f_A(x)$ if $x \in A$. Then f is well-defined and $f|_A = f_A$ for each $A \in \alpha$. Since X is an α_R-space, f is continuous on all of X. Clearly (f_i) converges to f. ∎

Corollary 5.1.2. Let X be any space.

(a) $C_k(X)$ is uniformly complete if and only if X is a k_R-space.

(b) $C_p(X)$ is uniformly complete if and only if X is discrete.

2. Čech-completeness and complete metrizability. A Tychonoff space which is a G_δ-subset of any Hausdorff space containing it as a dense subspace is called a Čech-complete space. Every completely metrizable space is Čech-complete, and conversely, every Čech-complete metrizable space is completely metrizable.

Theorem 5.2.1. The following are equivalent.

(a) $C_\alpha(X)$ is Čech-complete.

(b) $C_\alpha(X)$ is completely metrizable.

(c) $\alpha a(X) = \omega$ and X is an α_R-space.

Proof. Parts (a) and (b) are equivalent by Theorem 4.4.2, since every Čech-complete space is a q-space. That (b) implies (c) follows from Theorems 4.4.2 and 5.1.1 (also see Exercise 2(b)).

It remains to show that (c) implies (b). So let X be an α_R-space such that $\alpha a(X) = \omega$. Let $\{A_n: n \in \omega\} \subset \alpha$ be an α-cover for X, where each $A_n \subset A_{n+1}$.

The first goal is to show that a subset S of X is closed if and only if $S \cap A_n$ is closed in A_n for each n. So let S be a subset of X such that $S \cap A_n$ is closed for each n. Suppose, by way of contradiction, that S is not closed. Then S has an accumulation point x which is not in S. Since x is in some A_n, there is no loss of generality in assuming that $x \in A_1$. There exists a continuous $f_1: A_1 \to \mathbb{R}$ such that $f_1(S \cap A_1) = \{0\}$ and $f_1(x) = 1$. Now f_1 can be extended to a continuous $f_2: A_2 \to \mathbb{R}$ such that $f_2(S \cap A_2) = \{0\}$, and f_2 can be extended to f_3, and so on. Define function $f: X \to \mathbb{R}$ by $f(y) = f_n(y)$ for each $y \in A_n$. Since each $A \in \alpha$ is contained is some A_n, then f is continuous on each member of α, so that f would be continuous. But since $f(x) = 1$ while $f(S) = \{0\}$, f cannot be continuous. This contradiction establishes that S is closed.

For the final step, let Z be the topological sum of the A_n's, and let $p: Z \to X$ be the natural projection. The property in the previous paragraph shows that p is a quotient map. Therefore, the induced map $p^*: C_\alpha(X) \to C_\beta(Z)$ (where $\beta = \{A_n \cap A: A \in \alpha$ and $n \in \omega\}$) is a closed embedding. Since each $C_\alpha(A_n)$ is completely metrizable and since $C_\beta(Z)$ is homeomorphic to the product of the $C_\alpha(A_n)$, then $C_\alpha(X)$ is completely metrizable. ■

The proof of Theorem 5.2.1 shows, as a special case, that every hemicompact $k_{\mathbb{R}}$-space is a k-space.

<u>Corollary 5.2.2.</u> The following are equivalent.

(a) $C_k(X)$ is Čech-complete.

(b) $C_k(X)$ is completely metrizable.

(c) X is a hemicompact k-space.

<u>Corollary 5.2.3.</u> The following are equivalent.

(a) $C_p(X)$ is Čech-complete.

(b) $C_p(X)$ is completely metrizable.

(c) X is countable and discrete.

A completely metrizable space which is also separable is called a <u>Polish</u> space. Therefore Theorems 4.5.2 and 5.2.1 can be combined to characterize $C_\alpha(X)$ being a Polish space.

<u>Theorem 5.2.4.</u> The space $C_\alpha(X)$ is a Polish space if and only if $\alpha\alpha nw(X) = \omega$ and X is an α_R-space.

<u>Corollary 5.2.5.</u> Let X be any space.

(a) $C_k(X)$ is a Polish space if and only if X is a hemicompact k-space which has a countable network.

(b) $C_p(X)$ is a Polish space if and only if X is a countable discrete space.

<u>3. Baire Spaces.</u> A space is a <u>Baire space</u> provided that every countable intersection of open dense subsets is dense. This is implied by the other completeness properties. A homogeneous space, such as $C_\alpha(X)$, is a Baire space if and only if it is of second category in itself.

As before α is a hereditarily closed, compact network on X. Also assume that α is closed under finite unions. A good sufficient condition for $C_\alpha(X)$ to be a Baire space is hard to find. The next theorem gives a fairly strong sufficient condition in the case that α consists of the compact subsets of X.

<u>Theorems 5.3.1.</u> If X is a locally compact, paracompact space, then $C_k(X)$ is a Baire space.

Proof. A locally compact, paracompact space can be written as a topological sum of locally compact, σ-compact (and hence hemicompact) subspaces. So by Corollaries 2.4.7

and 5.2.2, $C_k(X)$ is homeomorphic to a product of completely metrizable spaces. Such a product is a Baire space. ■

The paracompactness in Theorem 5.3.1 cannot be omitted because the next theorem shows that if X is the space of countable ordinals, then $C_k(X)$ is not a Baire space. The local compactness in Theorem 5.3.1 also cannot be omitted since Theorem 5.3.3 shows that if X is the space of rational numbers, then $C_k(X)$ is not a Baire space.

Necessary conditions for $C_\alpha(X)$ to be a Baire space are easier to obtain, and generate many examples of function spaces which are of first category in themselves.

<u>Theorem</u> <u>5.3.2.</u> If $C_\alpha(X)$ is a Baire space, then every closed pseudocompact subset is contained in α. In particular, $C_\alpha(X) = C_k(X)$.

Proof. Suppose that A is a closed pseudocompact subset of X which is not contained in α, and hence not contained in any member of α. Then for each $n \in \omega$, the open set $G_n = \cup\{[a,(n,n+1)]: a \in A\}$ is dense in $C_\alpha(X)$. But since A is pseudocompact, $\cap\{G_n: n \in \omega\}$ is empty, which contradicts $C_\alpha(X)$ being Baire. ■

It follows that if $C_p(X)$ is a Baire space, then every closed pseudocompact subset of X is finite, and hence every compact subset is finite.

<u>Theorem</u> <u>5.3.3.</u> If $C_\alpha(X)$ is a Baire space, then each point in X which has a countable base has a neighborhood from α. In particular, if X is first countable, then it is locally compact.

Proof. Because of Theorem 5.3.2, it suffices to show that each point having a countable base has a compact neighborhood. Suppose that $x \in X$ has nested base $\{B_n: n \in \omega\}$, but that x does not have a compact neighborhood. For each n define $A_n =$

$\overline{B}_n \setminus B_{n+1}$. Suppose that each A_n were compact. Then let $\{x_n: n \in \omega\}$ be any sequence in $\overline{B}_1$. If $\{x_n: n \in \omega\}$ is contained in the union of finitely many of the A_n, then it would have a cluster point. Otherwise it would have a subsequence $\{x_{n_i}: i \in \omega\}$ such that each $x_{n_i} \in B$. Then by hypothesis $\{x_n: n \in \omega\}$ would still have a cluster point. But this would mean that $\overline{B}_1$ is compact, which is a contradiction. Therefore there is some $k_1 \in \omega$ such that A_{k_1} is not compact. By repeating this argument, an increasing sequence $\{k_n: n \in \omega\}$ can be found such that each A_{k_n} is not compact. Now for each n, define $G_n = \cup\{[a,(n,n+1)]: a \in A_{k_n}\}$, which is open and dense in $C_\alpha(X)$. Suppose there exists an $f \in \cap\{G_n: n \in \omega\}$. Then for each n, there exists an $a_n \in A_{k_n}$ such that $f(a_n) \in (n,n+1)$. But $\{a_n: n \in \omega\}$ would have a cluster point, which is impossible if f is continuous. Therefore $\cap\{G_n: n \in \omega\}$ is empty, which contradicts $C_\alpha(X)$ being a Baire space. ■

Therefore if $C_p(X)$ is a Baire space, then the only points of X which have countable bases are the isolated points.

Corollary 5.3.4. If X is a first countable paracompact space, then $C_k(X)$ is a Baire space if and only if X is locally compact.

The first countability hypothesis in Corollary 5.3.4 can be weakened to that of being a q-space.

A subfamily β of α will be said to move off α provided that for each $A \in \alpha$, there exists a $B \in \beta$ which is disjoint from A.

Theorem 5.3.5. If $C_\alpha(X)$ is a Baire space, then every subfamily from α which moves off α contains a countable discrete subfamily whose union is C-embedded in X.

Proof. Let β be a subfamily of α which moves off α. For each $n \in \omega$, define $G_n = \cup\{[A,(n,n+1)]: A \in \beta\}$, which is open in $C_\alpha(X)$. Each G_n is also dense in $C_\alpha(X)$ since β moves off α. Since $C_\alpha(X)$ is a Baire space, there exists an $f \in \cap\{G_n: n \in \omega\}$. So for each n, there exists an $A_n \in \beta$ such that $f \in [A_n,(n,n+1)]$. It is clear that $\{A_n: n \in \omega\}$ is a discrete family. So it remains to show that $\cup\{A_n: n \in \omega\}$ is C-embedded in X. Let $g: \cup\{A_n: n \in \omega\} \to \mathbb{R}$ be continuous. For each n, let $\phi_n: g(A_n) \to (n,n+1)$ be an embedding, let B_n be the boundary of $f^{-1}((n,n+1))$ and let C_n be the closure of $f^{-1}((n,n+1))$. Then for each n, define $\varsigma_n: C_n \to [n,n+1]$ so that it is continuous and so that $\varsigma_n|A_n = \phi_n \circ g|A_n$ and $\varsigma_n|B_n = f|B_n$. Now define $\varsigma: X \to \mathbb{R}$ by $\varsigma(x) = \varsigma_n(x)$ if $x \in C_n$ and $\varsigma(x) = f(x)$ if $x \in X \setminus \cup\{C_n: n \in \omega\}$, which is well-defined and continuous. Finally define $\psi: \mathbb{R} \to \mathbb{R}$ to be continuous such that for each n, $\psi|\phi_n \circ g(A_n) = \phi_n^{-1}|\phi_n \circ g(A_n)$. Then $\psi \circ \varsigma$ is a continuous extension of g to X, so that $\cup\{A_n: n \in \omega\}$ is C-embedded in X. ∎

A family of subsets of X is <u>strongly discrete</u> provided that the members of the family have neighborhoods which form a discrete family in X.

<u>Corollary 5.3.6.</u> If $C_\alpha(X)$ is a Baire space, then every sequence from α which moves off α has a strongly discrete subsequence.

The converse of this corollary is true, at least for the topology of pointwise convergence. This is shown in Theorem 5.3.8, which follows the next lemma.

<u>Lemma 5.3.7.</u> Let J be a closed interval in $\mathbb{R}$, let $f \in C(X,J)$, let A be a closed subset of X, let F be a finite subset of X, let $\epsilon > 0$, and let $t \in J^F$ such that $|t(x)-f(x)| < \epsilon$ for all $x \in F \cap A$. Then there exists a $g \in C(X,J)$ such that $g|F = t$ and $|g(x)-f(x)| < \epsilon$ for all $x \in A$.

Proof. Choose family $\{U_x : x \in F\}$ of pairwise disjoint open subsets of X such that $x \in U_x$ for each $x \in F$ and $U_x \subset X \setminus A$ for each $x \in F \setminus A$. Then for each $x \in F$, let J_x be the closed interval having 0 and $t(x)-f(x)$ as endpoints. For each such x, choose $\phi_x \in C(X,J_x)$ such that $\phi_x(x) = t(x)-f(x)$ and $\phi_x(X \setminus U_x) = \{0\}$. Define $\psi = f + (\Sigma\{\phi_x : x \in F\})$, and then modify ψ to get g as follows. Let $x \in X$. If $\psi(x)$ is greater than each element of J, then define g(x) to be the largest element of J. If $\psi(x)$ is less than each element of J, then define g(x) to be the smallest element of J. Otherwise take $g(x) = \psi(x)$. This defines the desired $g \in C(X,J)$. ∎

<u>Theorem</u> <u>5.3.8.</u> The space $C_p(X)$ is a Baire space if and only if every pairwise disjoint sequence of finite subsets of X has a strongly discrete subsequence.

Proof. The necessity follows from Corollary 5.3.6. The proof of the sufficiency makes strong use of the fact that $C_p(X)$ is a subspace of $\mathbf{R}^X$ with the product topology. Let $\mathcal{U}$ be the family of all basic open subsets of $\mathbf{R}^X$ of the form $\langle f,F,\epsilon\rangle = \{g \in \mathbf{R}^X : |f(x)-g(x)| < \epsilon$ for all $x \in F\}$, where $f \in \mathbf{R}^X$, F is a finite subset of X, and $\epsilon > 0$. For each $U = \langle f,F,\epsilon\rangle$ which is in $\mathcal{U}$, define $S(U) = F$ and $m(U) = \sup\{|g(x)| : g \in U$ and $x \in F\}$.

Now suppose that $\{F_n : n \in \omega\}$ is a sequence of nowhere dense subsets of $C_p(X)$, where each $F_n \subset F_{n+1}$. Since $C_p(X)$ is dense in $\mathbf{R}^X$, then each F_n is nowhere dense in $\mathbf{R}^X$.

Define three sequences by induction as follows. Increasing sequence $\{S_n : n \in \omega\}$ of finite subsets of X, increasing sequence $\{m_n : n \in \omega\}$ of positive numbers, and increasing sequence $\{\mathcal{U}_n : n \in \omega\}$ of finite subfamilies of $\mathcal{U}$ are to satisfy, for each n, the conditions:

(a) $U \cap F_n$ is empty for all $U \in \mathcal{U}_n$,

(b) $S(U) \subset S_n$ for all $U \in \mathcal{U}_n$,

(c) $f(S(U)) \subset [-m_n,m_n]$ for all $f \in U \in \mathcal{U}_n$,

(d) $(\cup\mathcal{U}_{n+1})$ intersects $<f,S_n,1/n>$ for all $f \in [-m_n,m_n]^X$.

To begin the induction, let $U \in \mathcal{U}$ such that $U \cap F_1$ is empty; then define $S_1 = S(U)$, $m_1 = m(U)$, and $\mathcal{U}_1 = \{U\}$. For the inductive step, suppose that S_n, m_n, and $\mathcal{U}_n$ have been defined satisfying (a) through (d). Let $Z = \{f \in [-m_n,m_n]^X : f(X \setminus S_n) = \{0\}\}$. Since Z is compact, there exist $f_1,...,f_k \in Z$ such that $Z \subset <f_1,S_n,1/2n> \cup ... \cup <f_k,S_n,1/2n>$. For each $i=1,...,k$, let $U_i \in \mathcal{U}$ such that $U_i \subset <f_i,S_n,1/2n> \setminus F_{n+1}$. Then define $S_{n+1} = S_n \cup S(U_1) \cup ... \cup S(U_k)$, $m_{n+1} = m_n + \max\{m(U_1),...,m(U_k)\}$, and $\mathcal{U}_{n+1} = \mathcal{U}_n \cup \{U_1,...,U_k\}$.

Conditions (a) through (c) are obviously satisfied by definition. To check that (d) is satisfied, let $f \in [-m_n,m_n]^X$. Then let $f_0 \in Z$ with $f_0|_{S_n} = f|_{S_n}$. Now there is an i such that $f_0 \in <f_i,S_n,1/2n>$. Let $g \in U_i$ and let $x \in S_n$. Then $|f(x)-g(x)| = |f_0(x)-g(x)| \le |f_0(x)-f_i(x)| + |f_i(x)-g(x)| < 1/2n + 1/2n = 1/n$. Therefore $U_i \subset <f,S_n,1/n>$, so that $(\cup\mathcal{U}_{n+1}) \cap <f,S_n,1/n>$ is nonempty.

The pairwise disjoint sequence $\{S_{n+1} \setminus S_n : n \in \omega\}$ has a strongly discrete subsequence $\{S_{n_k+1} \setminus S_{n_k} : k \in \omega\}$, where each $n_{k+1} \ge \max\{n_k+2,2k+1\}$. Relabel the terms of the above sequences as follows. For each $k \in \omega$, let $T_{2k-1} = S_{n_k}$, $T_{2k} = S_{n_k+1}$, $M_{2k-1} = m_{n_k}$, $M_{2k} = m_{n_k+1}$, $\mathcal{V}_{2k-1} = \mathcal{U}_{n_k}$, and $\mathcal{V}_{2k} = \mathcal{U}_{n_k+1}$.

These new sequences satisfy the following conditions for each n:

(1) $V \cap F_n$ is empty for all $V \in \mathcal{V}_n$,

(2) $S(V) \subset T_n$ for all $V \in \mathcal{V}_n$,

(3) $f(S(V)) \subset [-M_n,M_n]$ for all $f \in V \in \mathcal{V}_n$,

(4) $\cup\mathcal{V}_{n+1}$ intersects $<f,T_n,1/n>$ for all $f \in [-M_n,M_n]^X$.

To show this for (1) through (3), let $V \in \mathcal{V}_n$. Consider the case that $n = 2k$ for some $k \in \omega$; the case that $n = 2k-1$ is similar. Since $V \in \mathcal{V}_n = \mathcal{V}_{2k} = \mathcal{U}_{n_k+1}$, then $S(V) \subset S_{n_k+1} = T_{2k} = T_n$. Also for every $f \in V$, $f(S(V)) \subset [-m_{n_k+1},m_{n_k+1}] = [-M_{2k},M_{2k}] = [-M_n,M_n]$. It remains to show that $V \cap F_n$ is empty. Since $n_k \ge 2k-1$, then $n \le n_k+1$, so that $F_n \subset F_{n_k+1}$. It follows that $V \cap F_n$ is empty. To establish condition (4), let $f \in [-M_n,M_n]^X$, and again consider the case that $n = 2k$. Then there exists a $g \in (\cup \mathcal{U}_{n_k+2}) \cap \langle f, S_{n_k+1}, 1/(n_k+1)\rangle$.Since $n_{k+1} \geq n_k+2$, then $T_n = S_{2k} \subset S_{n_k+1}$. Finally since $1/(n_k+1) \leq 1/2k = 1/n$, then $g \in (\cup \mathcal{V}_{n+1}) \cap \langle f, T_n, 1/n\rangle$.

Let $\{W_{2k} : k \in \omega\}$ be a disjoint family of open subsets of X such that for each k:

(5) $T_{2k} \setminus T_{2k-1} \subset W_{2k}$,

(6) $T_{2k-1} \cap W_{2k}$ is empty.

Also make the following definition for each k:

(7) $D_{2k} = X \setminus \cup\{W_{2i} : i>k\}$.

Now define four additional sequences by induction as follows. Increasing sequence $\{j_n : n \in \omega\}$ of positive even integers, sequence $\{V_n : n \in \omega\}$ of members of $\mathcal{U}$, sequence $\{f_n : n \in \omega\}$ from $C_p(X)$, and sequence $\{\epsilon_n : n \in \omega\}$ of positive numbers are to satisfy, for each n, the conditions:

(8) $\epsilon_{n+1} \leq \epsilon_n/2$,

(9) $\langle f_n, T_{j_n}, 3\epsilon_n\rangle \subset V_n \in \mathcal{V}_{j_n}$,

(10) $f_n \in [-M_{j_n}, M_{j_n}]^X$,

(11) $|f_{n+1}(x) - f_n(x)| < \epsilon_n$ for all $x \in D_{j_n}$.

To begin the induction, let $j_1 = 2$, let $V_1 \in \mathcal{V}_2$, and let $f \in V_1$. By (3), $f(S(V_1)) \subset [-M_2, M_2]$. Then by Lemma 5.3.7, there is an $f_1 \in C_p(X)$ such that $f_1 \in [-M_2, M_2]^X$ and $f_1(x) = f(x)$ for all $x \in S(V_1)$. Then $f_1 \in V_1$, so that there exists an > 0 such that $\langle f_1, T_2, 3\epsilon_1\rangle$ ..n V_1.

For the induction step, suppose that j_n, V_n, f_n, and ϵ_n have been defined satisfying (8) through (11). Then let j_{n+1} be an even integer greater than $\max\{j_n, 1/_n + 1\}$. By (10) and (4), there exists a $V_{n+1} \in \mathcal{V}_{j_{n+1}}$ such that $V_{n+1} \cap \langle f_n, T_{j_{n+1}-1}, 1/(j_{n+1}-1)\rangle$ is nonempty; say it contains f. Since by (5) and (7), $T_{j_{n+1}} \setminus T_{j_{n+1}-1} \subset W_{j_{n+1}} \subset X \setminus D_{j_{n+1}-2} \subset X \setminus D_{j_n}$; since by (3), $f(S(V_{n+1})) \subset [-M_{j_{n+1}}, M_{j_{n+1}}]$ and since by (10), $f_n \in [-M_{j_n}, M_{j_n}]^X$; then Lemma 5.3.7 guarantees the existence of an $f_{n+1} \in C_p(X)$ such that $f_{n+1} \in [-M_{j_{n+1}}, M_{j_{n+1}}]^X$, $|f_{n+1}(x)-f_n(x)| < 1/(j_{n+1}-1) < \epsilon_n$ for all $x \in D_{j_n}$ and $f_{n+1}(x) = f(x)$ for all $x \in T_{j_{n+1}}$. Since by (2), $S(V_{n+1}) \subset T_{j_{n+1}}$, so that $f_{n+1} \in V_{n+1}$. Then there exists an $0 < \epsilon_{n+1} \le \epsilon_n/2$ such that $<f_{n+1}, T_{j_{n+1}}, 3\epsilon_{n+1}> \subset V_{n+1}$. Conditions (8) through (11) are now satisfied by these definitions.

Condition (11) can be modified to say:

(12) $|f_m(x)-f_n(x)| < 2\epsilon_k$ whenever $m,n \ge k$ and $x \in D_{j_k}$.

This follows from (11) and (8) since $|f_m(x)-f_n(x)| \le |f_m(x)-f_{m-1}(x)| + \ldots + |f_{n+1}(x)-f_n(x)| < \epsilon_{m-1} + \ldots + \epsilon_{n+1} + \epsilon_n \le \epsilon_n/(2^{m-n+1}) + \ldots + \epsilon_n/2 + \epsilon_n < 2\epsilon_n < 2\epsilon_k$.

In particular, (12) says that for each k, $\{f_n : n \in \omega\}$ is uniformly Cauchy on D_{j_k}. Therefore $\{f_n : n \in \omega\}$ converges pointwise to an $f \in \mathbb{R}^X$ such that f restricted to each D_{j_k} is continuous. Since $\{W_{2k} : k \in \omega\}$ is discrete, then $f \in C_p(X)$.

Finally, to show that $f \in V_n$ for each n, let n be fixed. Let $x \in S(V_n)$; so that by (2), (6) and (7), $x \in T_{j_n} \subset D_{j_n}$. If $m > n$, then by (12), $|f_m(x)-f_n(x)| < 2\epsilon_n$. This means that $|f(x)-f_n(x)| \le 2\epsilon_n < 3\epsilon_n$. By condition (9), then $f \in <f_n, T_{j_n}, 3\epsilon_n> \subset V_n$. Now letting n vary, by (1), $f \notin \cup\{F_{j_n} : n \in \omega\} = \cup\{F_n : n \in \omega\}$. It follows that $C_p(X)$ is of second category in itself, and is thus a Baire space. ■

4. An Application of Completeness. The proof of Theorem 5.4.2 in this section is an example of how the completeness of a function space can be used to show the existence of a certain kind of function, like an embedding.

The next theorem, which will be used in the proof of Theorem 5.4.2, might be called the Dugundji-Michael Extension Theorem. For a proof see Dugundji [1951] and Michael [1953].

Theorem 5.4.1. If A is a closed subspace of a metric space X and E is a locally convex linear topological space, then there is a linear embedding L: $C_k(A,E) \to C_k(K,E)$ such that $L(f)|_A = f$ for every $f \in C_k(A,E)$.

Actually the statement of Theorem 5.4.1 is not used in the proof of the next theorem so much as the method of proof itself.

<u>Theorem</u> <u>5.4.2.</u> Let X be a hemicompact metric space, let A be a closed subspace of X, and let E be an infinite-dimensional Banach space. If $h: A \to E$ is an embedding, then h has an extension $\hat{h}: X \to E$ which is also an embedding.

Proof. First consider the case that X is compact. Let d be a compatible metric on X and let $\|\cdot\|$ be the norm on E; then C(X,E) is to have the topology of uniform convergence with respect to this norm. Now C(X,E) is a complete metric space under this topology. Define $F = \{f \in C(X,E)$: f is an extension of h$\}$, which is a nonempty subspace of C(X,E) because of Theorem 5.4.1. It is straightforward to check that F is closed in C(X,E), so that F is a Baire space. For each natural number n, define $F_n = \{f \in F$: for each $y \in E$, the diameter of $f^{-1}(y)$ is less than $2/n\}$. If each F_n were open and dense in F, then since F is a Baire space, there would be an $\hat{h} \in \cap\{F_n: n \in \omega\}$. Such an $\hat{h}$ would necessarily be one-to-one, and would hence be the desired embedding since X is compact.

To show that F_n is open in F, let $f \in F_n$. Define $D = \{(x,y) \in X^2: d(x,y) \geq 1/n\}$, which is compact. Also define $\epsilon = \inf\{\|f(x)-f(y)\|: (x,y) \in D\}$. Since D is compact and since f and $\|\cdot\|$ are continuous, then $\epsilon > 0$. So let $g \in B(f,\epsilon/2)\cap F$. Suppose $x,y \in X$ such that $g(x) = g(y)$. Then $\|f(x)-f(y)\| \leq \|f(x)-g(x)\| + \|g(y)-f(y)\| < \epsilon$. Therefore $(x,y) \notin D$, so that $d(x,y) < 1/n$. It follows that $g \in F_n$, and hence F_n is open in F.

It remains to show that F_n is dense in F. So let $f \in F$ and $\epsilon > 0$. It is necessary to construct a g which is an element of $F_n \cap B(f,\epsilon)$. Since f is uniformly continuous on X, there exists a δ with $0 < \delta < 1/n$ such that whenever $d(x,y) < \delta$ then $\|f(x)-f(y)\| < \epsilon/4$. Now for each $x \in X \setminus A$, define $r_x = \min\{\delta, d(x,A)\}/2$. Then let $\mathcal{U} = \{U_m: m \in \omega\}$ be a countable locally finite open refinement of $\{B(x,r_x): x \in X \setminus A\}$ so that $\cup\mathcal{U} = X \setminus A$. For each m, choose an $x_m \in U_m$. Define a sequence (p_n) from E by induction as follows. First let $p_1 = f(x_1)$; then suppose $p_1,\ldots,p_n$ have been defined, and choose

$$p_{n+1} \in (E \setminus \mathrm{span}\{p_1,\ldots,p_n\}) \cap B(f(x_{n+1}),\varsigma),$$

where $\varsigma = \min\{\epsilon/4, d(x_{n+1},A)\}$. Now for each m, define the function $\lambda_m: X \setminus A \to \mathbb{R}$ by

$$\lambda_m(x) = \frac{d(x, X \setminus U_m)}{\Sigma\{d(x, X \setminus U_k): k \in \omega\}}.$$

Finally define the desired g: $X \to E$ by taking

$$g(x) = \Sigma\{\lambda_m(x) \cdot p_m: m \in \omega\}$$

if $x \in X \setminus A$ and $g(x) = f(x)$ for $x \in A$.

First establish that g is continuous. Because of the local finiteness of $\mathcal{U}$, g is clearly continuous at points of $X \setminus A$. So let $a \in A$ and $\tau > 0$. Since f is continuous at a, there exists a $\sigma > 0$ such that $\sigma < \tau$ and $f(B(a,\sigma)) \subset B(g(a),\tau/2)$. Then define $U = B(a,\sigma/4)$.

As an intermediate step, it must be shown that if U_m intersects U for some m, then U_m is contained in $B(a,\sigma)$. For such an m, there is an $x \in X \setminus A$ with $U_m \subset B(x,r_x)$; and hence there is a $y \in U \cap B(x,r_x)$. Then $d(x,a) \leq d(x,y) + d(y,a) < r_x + \sigma/4 \leq d(x,A)/2 + \sigma/4$. Since $d(x,y) < d(x,A)/2$, then $d(y,a) > d(x,A)/2$. So $d(x,A)/2 < \sigma/4$, and hence $d(x,a) < \sigma/2$. If $z \in B(x,r_x)$, then $d(z,x) < r_x \leq d(x,A)/2 < \sigma/4$. Therefore $d(z,a) \leq d(z,x) + d(x,a) < \sigma/4 + \sigma/2 < \sigma$. This means that $U_m \subset B(a,\sigma)$ as desired.

To complete the proof that g is continuous, it must be shown that $g(U) \subset B(g(a),\tau)$; so let $x \in U$. If $x \in A$, then $g(x),f(x) \in f(B(a,\sigma)) \subset B(g(a),\tau/2)$. On the other hand, consider $x \in X \setminus A$. Then for each m for which $x \in U_m$, U_m intersects U and is thus contained in $B(a,\sigma)$. Then for such m, $f(x_m) \in B(g(a),\tau/2)$ and $d(x_m,A) \leq$

$d(x_m,a) < \sigma/2 < \tau/2$. Therefore $\|p_m - g(a)\| \le \|p_m - f(x_m)\| + \|f(x_m) - g(a)\| < \tau$, so that $p_m \in B(g(a),\tau)$. It follows that $g(x) \in B(g(a),\tau)$ since $B(g(a),\tau)$ is convex. This finishes the proof that g is continuous.

The next thing to show is that $g \in B(f,\epsilon)$. Let $x \in X \setminus A$, and let $m_1,\ldots,m_k$ be those m such that $x \in U_m$. For each $i = 1,\ldots,k$, $\|f(x) - f(x_{m_i})\| < \epsilon/4$ and $\|f(x_{m_i}) - p_{m_i}\| < \epsilon/4$, so that $\|f(x) - p_{m_i}\| < \epsilon/2$. Therefore $\|f(x) - g(x)\| = \|\lambda_{m_1}(x)f(x) + \ldots + \lambda_{m_k}(x)f(x) - \lambda_{m_1}(x)p_{m_1} - \ldots - \lambda_{m_k}(x)p_{m_k}\| \le \lambda_{m_1}(x)\|f(x) - p_{m_1}\| + \ldots + \lambda_{m_k}(x)\|f(x) - p_{m_k}\| < \epsilon/2$, and hence $g \in B(f,\epsilon)$.

The final thing to show is that $g \in F_n$. Again let $x \in X \setminus A$, and let $m_1,\ldots m_k$ be those m such that $x \in U_m$. Now define $L(x)$ to be the linear span of $\{p_{m_1},\ldots,p_{m_k}\}$, so that $g(x) \in L(x)$. Suppose $y \in X \setminus A$ is such that $L(y)$ intersects $L(x)$; say $L(y)$ is spanned by $\{p_{q_1},\ldots,p_{q_l}\}$. Then one of the p_{m_i} must equal one of the p_{q_j}, so that both x and y are in U_m for some m. This means that if $g(x) = g(y)$ for some $y \in X \setminus A$, then there is an m such that $x,y \in U_m$. But then $d(x,y) < \delta$, and hence the diameter of $g^{-1}(x)$ is less than or equal to δ and thus less than $1/n$. It follows that $g \in F_n$, which concludes the proof in the case that X is compact.

For the general hemicompact case, X can be written as $\cup\{X_n : n \in \omega\}$, where each X_n is compact and contains the interior of X_{n+1}. Then a sequence of embeddings h_n: $X_n \to E$ can be defined by induction as follows. Let $g_0 = h|_{X_1 \cap A}$; and for the induction step let each h_n be the extension of g_{n-1} to A_n which is promised by the argument above, where each g_n: $X_n \cup (X_{n+1} \cap A) \to E$ is the combination function of g_{n-1} and $h|_{X_{n+1} \cap A}$. Now the combination function of all the h_n's is the desired embedding of X into E which extends h. ∎

The "hemicompact" hypothesis for X in Theorem 5.4.2 cannot be replaced with "separable", unless A is compact. This is because the embedding of A into E might fill up E with no room for an extension. For example, a closed ball in a separable infinite-dimensional Banach space E is homeomorphic to E itself.

5. Countably Compact Subsets. The main goal of this section is to establish the Grothendieck Theorem which asserts that every countably compact subset of $C_p(X,R)$ is compact whenever X is compact and R is metrizable. This is done by stages. The first stage is an interesting result in its own right.

Theorem 5.5.1. Let X be a compact space, and let R be a metric space. If F is a compact separable subspace of $C_p(X,R)$, then F is metrizable.

Proof. Since a compact space having a countable network is metrizable, it suffices to show that F has a countable network. Let Y be X with the weak topology with respect to F, and let $\mathcal{B} = \{f_1^{-1}(V_1) \cap \ldots \cap f_n^{-1}(V_n): f_1,\ldots,f_n \in F \text{ and } V_1,\ldots,V_n \text{ are open in } R\}$ be the natural base for Y. Since the identity map from X to Y is continuous and onto, then its induced function is an embedding from $C_p(Y,R)$ into $C_p(X,R)$. Each member of F is continuous on Y, so F may be considered as a subspace of $C_p(Y,R)$.

Define $\Delta: Y \to R^F$ to be the diagonal function. Since F separates points from closed sets, Δ is a continuous function which is open onto its image. Also for each $U \in \mathcal{B}$, $\Delta^{-1}(\Delta(U)) = U$ (although Δ need not be one-to-one). Then it is easy to see that if $\Delta(Y)$ has a countable base, so does Y. Let D be a countable dense subset of F, and let $p: R^F \to R^D$ be the natural projection. Since D is dense in F, $p|_{\Delta(Y)}$ is one-to-one. Since X is compact, so is Y and so is $\Delta(Y)$. Thus $p|_{\Delta(Y)}$ is a homeomorphism. Since R^D is metrizable, then so is $\Delta(Y)$. Therefore $\Delta(Y)$ has a countable base, so that Y has a countable base. Then by Exercise IV.9.11.a, $C_p(Y,R)$ has a σ-discrete network. This means that F has a σ-discrete network. But since F is compact, then this network is in fact countable. ∎

Corollary 5.5.2. Let X be a compact space, and let R be a metric space. Then every compact subspace of $C_p(X,R)$ is a Fréchet space.

Proof. Let F be a compact subspace of $C_p(X,R)$, let G be a subset of F and let g be an accumulation point of G in F. As indicated in the paragraph following Corollary 4.7.3, $C_p(X,R)$ has countable tightness. Since this is a hereditary property, there exists a countable subset C of G so that g is an accumulation point of C. Let D be the closure of C in F. Then D is a compact separable subspace of $C_p(X,R)$, and is hence metrizable by Theorem 5.5.1. Therefore there exsists a sequence in C which converges to g. ■

Lemma 5.5.3. Let X be a space, and let R be a metric space. If F is a countably compact subset of $C_p(X,R)$, then the closure of F in R^X is compact.

Proof. For each $x \in X$, let e_x: $C_p(X,R) \to R$ be the evaluation map defined by $e_x(f) = f(x)$. Since each e_x is continuous, then each $e_x(F)$ is a countably compact subset of R, and is therefore compact. So the closure of F in R^X is contained in $\Pi\{e_x(F): x \in X\}$, which is compact. ■

Lemma 5.5.4. Let X be a compact space, and let R be a compact metric space. If F is a countably compact subset of $C_p(X,R)$, then the closure of F in R^X is contained in $C_p(X,R)$.

Proof. Suppose, by way of contradiction, that there is an accumulation point f of F in R^X which is not continuous at some x_0 in X. Then $f(x_0)$ has a neighborhood V such that x_0 is not an interior point of $f^{-1}(V)$. By induction, define sequences (f_n) in F and (x_n) in X such that for each n,

$$f_n \in F \cap [x_0,B(f(x_0),1/n)] \cap \ldots \cap[x_{n-1},B(f(x_{n-1}),1/n)],$$

and

$$x_n \in (f_1^{-1}(B(f_1(x_0),1/n)) \cap \ldots \cap f_n^{-1}(B(f_n(x_n),1/n))) \setminus f^{-1}(V).$$

Since F is countably compact, F contains a cluster point g of (f_n). Let (x_{n_i}) be a subsequence of (x_n) such that $(f(x_{n_i}))$ converges to some r in R. Also let x be a cluster point of (x_{n_i}) in X. For the moment fix n. By the construction of the x_{n_i}, $(f(x_{n_i}))_i$ converges to some $f_n(x_0)$. But since x is a cluster point of (x_{n_i}) and f_n is continuous, then $f_n(x) = f_n(x_0)$. Now letting n vary, by the construction of the f_n, $(f_n(x_0))$ converges to $f(x_0)$. Also since g is a cluster point of (f_n) and since $f_n(x) = f_n(x_0)$, then $g(x) = f(x_0)$. For fixed i, use the same argument to show that $(f_n(x_{n_i}))_n$ converges to $f(x_{n_i})$ and that $g(x_{n_i}) = f(x_{n_i})$. Also for each i, $f(x_{n_i}) \notin V$, so that each $g(x_{n_i}) \notin V$. On the other hand, $g(x) = f(x_0) \in V$. Then since x is a cluster pont of (x_{n_i}), g cannot be continuous at x. But this contradicts g being in F. ∎

<u>Lemma 5.5.5.</u> Let X be a compact space, and let R be a metric space. If F is a countably compact subset of $C_p(X,R)$, then the closure of F in R^X is contained in $C_p(X,R)$.

Proof. Let f be an accumulation point of F in R^X, let $x \in X$ and let (x_i) be a net in X converging to x. Assume that ρ is a compatible metric on R which is bounded by 1. Define continuous ϕ: $R \to [0,1]$ by $\phi(t) = \rho(t,f(x))$ for each $t \in R$. Then the induced function ϕ_*: $R^X \to [0,1]^X$ is continuous and $\phi_*(C_p(X,R)) \subset C_p(X,[0,1])$. Thus $\phi_*(F)$ is a countably compact subset of $C_p(X,[0,1])$. By Lemma 5.5.4 it follows that the closure of $\phi_*(F)$ in $[0,1]^X$ is contained in $C_p(X,[0,1])$. Therefore $\phi_*(f)$ must be continuous. But then $(\phi(f(x_i))) = (\phi_*(f)(x_i))$ is a net converging to $\phi_*(f)(x) = \phi(f(x))$. That is, the net $(\rho(f(x_i), f(x)))$ converges to $\rho(f(x),f(x)) = 0$, so that $(f(x_i))$ converges to $f(x)$. It follows that f is continuous. ∎

<u>Theorem 5.5.6.</u> Let X be compact, let R be metrizable, and let F be a subspace of $C_p(X,R)$. Then the following are equivalent.

(a) F is compact.

(b) F is sequentially compact.

(c) F is countably compact.

Proof. To establish that (c) implies (a), let F be countably compact. By Lemma 5.5.5, the closure of F in R^X is contained in $C_p(X,R)$; and by Lemma 5.5.3, this closure is compact. It therefore suffices to show that F is closed in $C_p(X,R)$. So let f be an accumulation point of F in $C_p(X,R)$. By Corollary 5.5.2, there exists a sequence (f_n) in F which converges to g. Since F is countably compact, then (f_n) must have a cluster point in F, which must be equal to f.

Finally, to show that (a) implies (b), let F be compact, and let (g_n) be a sequence in F. By Corollary 5.5.2, it follows that (g_n) has a subsequence which converges to g. ∎

This theroem can be generalized to those spaces X which contain dense σ-compact subsets (cf. Exercise 8). It is also true for any (regular) topology on C(X,R) which is finer than the topology of pointwise convergence. A simple proof of this theorem for the compact-open topology is to use Corollary 4.3.2.

__6.__ __Dense Subsets.__ Throughout this section, the range space is the space $\mathbb{R}$ of real numbers, and $C_k(X,\mathbb{R})$ is abbreviated as $C_k(X)$. This space is a locally convex linear topological space. In fact not only are addition and scalar multiplication in $C_k(X)$ continuous, but so is multiplication of functions, so that $C_k(X)$ is a topological algebra. A subset of $C_k(X)$ is then called a __subalgebra__ provided that it is closed under addition, multiplication, and scalar multiplication.

The following well-known theorem is called the Stone-Weierstrass Theorem. A proof can be found for example in Dugundji [1968].

__Theorem 5.6.1.__ Every subalgebra of $C_k(X)$ which contains a nonzero constant function and separates points is dense in $C_k(X)$.

The Stone-Weierstrass Theorem and the Ascoli Theorem can be used together to characterize those spaces $C_k(X)$ which contain dense σ-compact subsets.

Theorem 5.6.2. If X is submetrizable, then $C_k(X)$ contains a dense σ-compact subset.

Proof. Let ϕ: X $\to$ M be a continuous injection into a metric space M. Since the induced function ϕ^*: $C_k(M) \to C_k(X)$ is continuous and maps onto a dense subspace of $C_k(X)$, then if $C_k(M)$ has a dense σ-compact subset, so does $C_k(X)$. Therefore it suffices to assume that X is a metric space.

A metric space is paracompact, so for each $n \in \omega$, there is a locally finite partition of unity F_n contained in $C_k(X)$ which is subordinated to the family of all open sets of diameter less than $1/n$. That is, F_n satisfies:

(a) for each $f \in F_n$, $f(x) \subset [0,1]$,

(b) $\{f^{-1}((0,1]): f \in F_n\}$ is locally finite,

(c) for each $f \in F_n$, the diameter of $f^{-1}((0,1])$ is less than $1/n$, and

(d) for each $x \in X$, $\Sigma\{f(x): f \in F_n\} = 1$.

Let e be the constant function mapping X to 1. For each $n \in \omega$, define $F_n' = \{rf: r \in [-1,1] \text{ and } f \in F_n \cup \{e\}\}$. To show that F_n' is evenly continuous, let $x \in X$ and let $V = (t-\epsilon,t+\epsilon)$ in $\mathbb{R}$. There exists a neighborhood U_0 of x and finite subset F of F_n such that for each $f \in F_n \setminus F$, $U_0 \cap f^{-1}((0,1])$ is empty. For each $f \in F$, there exists a neighborhood U_f of x such that $f(U) \subset (f(x)-\epsilon/2, f(x)+\epsilon/2)$. Define $U = U_0 \cap (\cap\{U_f: f \in F\})$ and $V' = (t-\epsilon/2, t+\epsilon/2)$. For each $f \in F$ and $r \in [-1,1]$, if $rf(x) \in V'$ then $rf(U) \subset V$. Also for each $f \in (F_n \cup \{e\}) \setminus F$ and $r \in [-1,1]$, rf is constant on U, so that in any case $rf(U) \subset V$ whenever $rf(x) \in V'$.

For each n, define $G_n = \{f_1 \ldots f_k: k=1,\ldots,n \text{ and } f_1,\ldots,f_k \in F_1' \cup \ldots \cup F_n'\}$, and define $H_n = \{f_1+\ldots+f_k: k=1,\ldots,n \text{ and } f_1,\ldots,f_k \in G_n\}$. Since $F_1' \cup \ldots \cup F_n'$ is evenly continuous, then G_n is evenly continuous, and thus so is H_n (cf. Exercise III.3.4). Since each H_n is clearly pointwise bounded, then by the Ascoli Theorem it has compact closure in $C_k(X)$.

Finally define $H = \cup\{H_n: n \in \omega\}$.

To show that H is a subalgebra of C(X), let $f,g \in H$; say $f \in H_m$ and $g \in H_n$. Then $f = f_1+...+f_k$ and $g = g_1+...+g_\ell$ where each $f_i \in G_m$ and each $g_i \in G_n$. Each f_i and each g_i is a product of m+n or fewer elements of $F_1' \cup ... \cup F_{m+n}'$, so that $f+g \in H_{m+n} \subset H$. A similar argument shows that $fg \in H_{mn} \subset H$. Finally, let $t \in \mathbb{R}$ and let $f \in H_m$. There is some n so that $|t| \leq n$. Then $nf \in H$, so that $tf = (t/n)nf \in H_{mn}$. Therefore H is a subalgebra of C(X).

It remains to show that H separates points. Let $x,y \in X$ with $x \neq y$. Then there exists an $n \in \omega$ so that the distance between x and y is greater than 1/n. Since F_n is a partition of unity, there is some $f \in F_n$ such that $f(x) \neq 0$. But then f is supported on B(x,1/n), so that $f(y) = 0$. So H separates points, and therefore H is dense in $C_k(X)$ by the Stone-Weierstrass Theorem. It follows that $C_k(X)$ contains a dense σ-compact subset. ∎

Whenever X is a k-space,the converse of Theorem 5.6.2 is true.

<u>Theorem</u> <u>5.6.3.</u> Let X be a k-space. Then $C_k(X)$ contains a dense σ-compact subset if and only if X is submetrizable.

Proof. If $C_k(X)$ contains a dense σ-compact subset, then $C_k(C_k(X))$ is submetrizable by Corollary 4.3.2. Since X is a k-space, the diagonal function $\Delta: X \to C_k(C_k(X))$ is an embedding by Theorem 2.3.6. Therefore X must be submetrizable. ∎

It is an interesting fact that Theorem 5.6.3 is the dual to the statement in Corollary 4.3.2 which says that $C_k(X)$ is submetrizable if and only if X contains a dense σ-compact subset.

The corresponding duality for the topology of pointwise convergence is contained in Corollaries 4.2.2 and 4.3.3. In particular, $C_p(X)$ is separable if and only if X contains a coarser separable metrizable topology; and $C_p(X)$ contains a coarser separable metrizable topology if and only if X is separable.

In fact these dualities can be put together to obtain the following.

Theorem 5.6.4. Let α be any compact network on X. Then $C_\alpha(X)$ is separable and submetrizable if and only if X is separable and submetrizable.

Up to this point, this section has been concerned with dense subsets of $C_\alpha(X)$. However $C_\alpha(X)$ may itself be considered as a dense subset of the larger space $\mathbb{R}^X$ with an appropriate topology. In particular, $C_p(X)$ is dense in $\mathbb{R}^X$ with the product topology. One can ask what kind of subset of $\mathbb{R}^X$ is $C_p(X)$. For example, when is $C_p(X)$ a G_δ-subset of $\mathbb{R}^X$ or an F_σ-subset of $\mathbb{R}^X$; or more generally, when is $C_p(X)$ a Borel subset of $\mathbb{R}^X$? The simplest such result is the following (cf. Exercise 9).

Theorem 5.6.5. The space $C_p(X)$ contains a dense G_δ-subset of $\mathbb{R}^X$ if and only if X is discrete (in which case $C_p(X) = \mathbb{R}^X$).

Proof. Suppose that X is not discrete. Then there exists a function $f \in \mathbb{R}^X \setminus C_p(X)$. Define the shifting function θ_f: $\mathbb{R}^X \to \mathbb{R}^X$ by $\theta_f(g) = f+g$ for every $g \in \mathbb{R}^X$. This is a homeomorphism such that $\theta_f(C_p(X)) \subset \mathbb{R}^X \setminus C_p(X)$. Since $\mathbb{R}^X$ is a Baire space, then $C_p(X)$ cannot contain a dense G_δ-subset since this would cause $\mathbb{R}^X \setminus C_p(X)$ to also contain a dense G_δ-subset, disjoint from that in $C_p(X)$. ∎

There is a related result when the denseness requirement is removed.

Theorem 5.6.6. The space $C_p(X)$ contains a nonempty G_δ-subset of $\mathbb{R}^X$ if and only if X is the topological sum of a countable space and a discrete space.

Proof. Let $f \in \cap\{W_n: n \in \omega\} \subset C_p(X)$, where each $W_n = \{g \in \mathbb{R}^X: |g(x)-f(x)| < \epsilon_n$ for each $x \in X_n\}$ for some $\epsilon_n > 0$ and finite subset X_n of X. Define $Y = \cup\{X_n: n \in \omega\}$ and $Z = X \setminus Y$. Note that if g $\mathbb{R}^X$ with $g|_Y = f|_Y$, then $g \in C_p(X)$.

To show that Y contains all the accumulation points of X, suppose there is an accumulation point x of X which is contained in Z. Then define $g: X \to \mathbb{R}$ by $g(y) = f(y)$ if $y \in X \setminus \{x\}$ and $g(x) = f(x)+1$. Now g is continuous since $g|_Y = f|_Y$. But since $X \setminus \{x\}$ is dense in Hausdorff space X, then f = g, which is a contradiction. Therefore Z is open and discrete in X.

Finally to show that Z has no accumulation point in X, again suppose that there is an accumulation point x of Z (which is contained in Y). Define $g: X \to \mathbb{R}$ by $g(y) = f(y)$ if $y \in Y$ and $g(y) = f(x)+1$ if $y \in Z$. Again g is continuous since $g|_Y = f|_Y$. Since x is an accumulation point of Z and g is constant on Z, then $g(x) = f(x)+1$. But since $x \in Y$, this is a contradiction. Therefore Z is closed in X.

For the converse, suppose X is the topological sum of Y and Z, where Y is countable and Z is discrete. Then $C_p(X)$ is homeomorphic to $C_p(X) \times \mathbb{R}^Z$, which is contained in $\mathbb{R}^X$. Let $f \in C_p(X)$. Since $\mathbb{R}^Y$ is metrizable, then f is a G_δ-subset of $\mathbb{R}^Y$. Therefore $\{f\} \times \mathbb{R}^Z$ is a nonempty G_δ-subset of $\mathbb{R}^X$ which is contained in $C_p(Y) \times \mathbb{R}^Z$. ■

7. Analytic Spaces. In Theorem 5.2.4, a characterization is given for a function space to be a Polish space. Now in this section a characterization is given for a function space to be a continuous image of a Polish space. A Tychonoff space is an analytic (or a Souslin) space if it is a continuous image of a Polish space, or, equivalently a continuous image of the space ω^ω (which is homeomorphic to the space of irrationals with the usual topology). In order to characterize this property on the function space, the concept of a Borel structure on a set is needed.

Let X be a non-empty set. A Borel structure on X is any non-empty collection $\mathcal{B}$ of subsets of X that is closed under taking complements and countable unions. This should be compared with the concept of a topological structure (i.e. topology) on a set X.

If $\mathcal{B}$ is a Borel structure on X then X together with $\mathcal{B}$, denoted by $(X,\mathcal{B})$ is called a Borel (or a measurable) space. Members of $\mathcal{B}$ are named Borel sets. If (X,τ) is a topological space, the smallest (in the sense of inclusion) Borel structure on X containing all the open sets of (X,τ) is called the Borel structure generated by (X,τ). This structure is denoted by $\mathcal{B}_\tau$. Borel structures on subsets and products are defined similarly to their topological counterparts. If $(X,\mathcal{B})$ and $(y,\mathcal{D})$ are two Borel spaces, a map $f: (X,\mathcal{B}) \to (Y,\mathcal{D})$ is measurable if and only if for each $D \in \mathcal{D}$, $f^{-1}(D) \in \mathcal{B}$. If f is one-to-one and onto and if f and f^{-1} are both measurable, f is called a Borel isomorphism. If $\mathcal{B}$ and $\mathcal{B}'$ are two Borel structures on X, $\mathcal{B}$ is isomorphic to $\mathcal{B}'$ means that the identity map id: $(X,\mathcal{B}) \to (X,\mathcal{B}')$ is a Borel isomorphism. A Borel space $(X,\mathcal{B})$ is <u>analytic</u> if $\mathcal{B}$ is isomorphic to $\mathcal{B}_\tau$ for some analytic topology τ on X. In the sequel, whenever X is called an analytic space with no mention of the structure, it means that X is analytic as a topological space. It is clear that the Borel space $(X,\mathcal{B}_\tau)$ is analytic whenever the topology τ is analytic. The converse however is not true as the Sorgerfrey line is a counterexample. But if τ is a metrizable topology on X then (X,τ) is analytic if and only if $(X,\mathcal{B}_\tau)$ is analytic. For more results on Borel structures, see Christensen [1974].

<u>Theorem</u> <u>5.7.1.</u> If X is a k_R-space then there exists no continuous surjection from X onto $C_k(X)$.

Proof. Suppose by way of contradiction that there is a continuous surjection φ: X $\to C_k(X)$. Define $f: X \to \mathbb{R}$ by $f(x) = 1 + \varphi(x)(x)$ for each x in X. Because of Corollary 2.5.4.a, for each compact A in X, $f|_A$ is continuous. It follows that f is continuous on all of X, since X is a k_R-space. Now let $x \in X$ such that $\varphi(x) = f$. Then $1 + \varphi(x)(x) = f(x) = \varphi(x)(x)$, which is impossible. ∎

The next result follows from Theorem 5.7.1, the definition of an analytic space, and the fact that ω^ω is a k_R-space.

<u>Corollary 5.7.2.</u> $C_k(\omega^\omega)$ is not an analytic space.

Since $\mathbb{R}$ is homeomorphic to $(0,+\infty)$ then $C_k(X)$ is clearly homeomorphic to its subspace $C_k^+(X) = \{f \in C(X) : f(X) \subset (0,+\infty)\}$, and $\mathbb{R}^X$ is homeomorphic to its subspace $\mathbb{R}_+^X = \{f:X \to \mathbb{R} : f(X) \subset (0,+\infty)\}$. Also for a space X, put $K(X) = \{A : A$ is a non-empty compact subset of $X\}$ and $\mathbb{R}_+^X = \{f \in \mathbb{R}^X : f(X) \subset (+\infty)\}$. The following characterization of Polish spaces can be found in Christensen [1974].

<u>Lemma 5.7.3.</u> A separable metrizable space Y is a Polish space if and only if there exists a Polish space X and a map $p: K(X) \to K(Y)$ satisfying:

(1) if $A,B \in K(X)$ with $A \subset B$, then $p(A) \subset p(B)$; and

(2) if $C \in K(Y)$, then there exists an $A \in K(X)$ such that $C \subset p(A)$.

The next theorem is the main result of this section.

<u>Theorem 5.7.4.</u> Let X be a k-space. Then $C_k(X)$ is an analytic space if and only if X is a σ-compact space having a countable k-netweight.

Proof. Suppose that $C_k(X)$ is analytic. Then $C_k(X)$ has countable netweight (since analyticity is stronger then the property of being a continuous image of a separable metric space). By Corollary 4.1.3, X has countable netweight. Let τ be the topology on X, and choose a coarser separable metrizable topology τ' on X with the following properties:

(1) (X,τ) and (X,τ') have the same Borel structure,

(2) (X,τ) is analytic if and only if (X,τ') is analytic.

The first step is to show that (X,τ') is analytic. To this end, let (Y,ρ) be a compact metric space (with metric ρ) in which (X,τ') is densely embedded and let Z =

$(Y,\rho)\setminus(X,\tau')$. Then Z is separable metrizable. To see that Z is a Polish space, let first $\varphi\colon \omega^\omega \to C_k(X)$ be a continuous surjection. Then define $\Gamma\colon \omega^\omega \to \mathbb{R}_+^X$ by $\Gamma(\sigma)(x) = \inf\{\varphi(\beta)(x) : \beta \le \sigma\}$ for all x in X and each $\sigma \in \omega^\omega$; where $\beta = (\beta(1),\beta(2),\ldots) \le \sigma = (\sigma(1),\sigma(2),\ldots)$ means that $\beta(n) \le \sigma(n)$ for each $n \in \omega$. One checks easily that if $\beta \le \sigma$ in ω^ω then $\Gamma(\beta) \ge \Gamma(\sigma)$ and if $f \in C^+(X)$ then $f \ge \Gamma(\sigma)$ for some $\sigma \in \omega^\omega$. Now, define $\psi\colon \omega^\omega \to K(Y)$ by $\psi(\sigma) = \cap\{Y\setminus B_x : x \in X\}$ where $B_x = \{y \in Y : \rho(x,y) < \Gamma(\sigma)(x)\}$ is the open ball about x of radius $\Gamma(\sigma)(x)$. Next, if A is compact in Z, the continuous map $f\colon Y \to \mathbb{R}$ defined by $f(y) = \rho(A,y)$ for all y in Y has its restriction on X in $C^+(X)$ and, by construction of Γ, $f|_X \ge \Gamma(\sigma)$ for some σ in ω^ω. Clearly then $A \subset \psi(\sigma)$.

Finally the map $p\colon K(\omega^\omega) \to K(Z)$ given by $p(\{\beta : \beta \le \sigma\}) = \psi(\sigma)$ satisfies the properties of Lemma 5.7.3. This shows that Z is a Polish space. This implies that Z is a G_δ-subspace of (Y,ρ), so that $(Y,\rho)\setminus Z = (X,\tau')$ is an F_σ-subspace of the compact metric space (Y,ρ). It follows that (X,τ') is a σ-compact metrizable space. Therefore (X,τ') is analytic, being a countable union of compact metric spaces. Now (X,τ') is analytic if and only (X,τ) is analytic.

The next step is to show that (X,τ) is σ-compact. It is known that an analytic space is σ-compact if and only if it does not contain a copy of ω^ω. So, by way of contradiction, assume that (X,τ) contains a closed copy of ω^ω and let $i\colon \omega^\omega \to (X,\tau)$ be the inclusion map. Then $i^*\colon C_k(X,\tau) \to C_k(\omega^\omega)$ is a continuous surjection. Since $C_k(X,\tau)$ is analytic, then $C_k(\omega^\omega)$ is also analytic, which contradicts Corollary 5.7.2. Therefore, (X,τ) must be a σ-compact space. This finishes the proof of the necessity.

For the sufficiency, suppose that $X = A_1 \cup A_2 \cup \ldots$, where each A_n is compact and X has countable k-netweight. Then there are separable metric space (M,ρ) with metric ρ and compact-covering map $\varphi\colon (M,\rho) \to X$. Since X is a k-space, φ is a quotient map so that $\varphi^*\colon C_k(X) \to C_k(M)$ is a closed embedding of $C_k(X)$ into $C_k(M)$ (see Corollary 2.2.8 and Theorem 2.2.10).

Now it remains to show that $\varphi^*(C_k(X))$ is a subspace of some analytic subspace of $C_k(M)$. By Theorem 4.3.4, there is a metric ρ' on X inducing a coarser separable metrizable topology. There is no loss of generality to assume that the metric ρ on M is such that the composite function $(M,\rho)\xrightarrow{\varphi}X\xrightarrow{id}(X,\rho')$ is uniformly continuous. Now, let $W = \{g \in C_k(M) : g$ is uniformly continuous on each $\varphi^{-1}(A_n)\}$. Note that for each $g \in C(X)$, $\varphi^*(g) = g\circ\varphi$ is uniformly continuous on each $\varphi^{-1}(A_n)$; so that $\varphi^*(C_k(X)) \subset W$. The claim is that W is an analytic subspace of $C_k(M)$.

The first step is to show that W is Borel isomorphic to an analytic space. To see this, let $\{B_1,B_2,B_3,...\}$ be a countable base for M, and choose $s_n \in B_n$ for each n. Then define ψ: $C_k(M) \to \mathbb{R}^S$ by $\psi(g) = \langle g(s_1),g(s_2),...\rangle$ for each g in C(M); where $S = \{s_1,s_2,s_3,...\}$. Then it can be shown that ψ is a Borel isomorphism of $C_k(M)$ onto its image $\psi(C_k(M))$. Moreover, $\psi(W)$ is a Borel set in the Polish space $\mathbb{R}^S$. It follows that $\psi(W)$ is an analytic subspace of $\psi(C_k(M))$, and that $\psi|_W$: $W \to \psi(W)$ is a Borel isomorphism.

Next, choose $\{r_1,r_2,...\}$ as a dense sequence in $\mathbb{R}$, and define d: $C(M)\times C(M) \to \mathbb{R}$ by

$$d(f,g) = \sum_{k,n} \frac{1}{2^{k+n}}\left|(\sup_{x\in B_n}|f(x)-r_k|\wedge 1)-(\sup_{x\in B_n}|g(x)-r_k|\wedge 1)\right|$$

for each f and g in C(M). Then d is a metric on C(M) such that id: $C_d(M) \to C_k(M)$ is continuous. Moreover the metric space $C_d(M)$ and the spaces $C_k(M)$ and $C_p(M)$ all have the same Borel structure. If we denote by (W,d) the set W as a subspace of the metric space $C_d(M)$, then W and (W,d) have the same Borel structure. It follows that the metric space (W,d) is in fact an analytic topological space. Since the identity map id: $(W,d) \to W \subset C_k(M)$ is continuous, then clearly W becomes an analytic topological subspace of $C_k(M)$. It then follows that its closed subspace $\psi^*(C_k(X))$ is analytic. $C_k(X)$ is therefore analytic. ∎

For a q-space X, Theorem 5.7.4 can be improved as follows.

<u>Theorem</u> <u>5.7.5.</u> If X is a q-space then the following are equivalent.

(1) $C_k(X)$ is an analytic space.

(2) $C_p(X)$ is an analytic space.

(3) X is a σ-compact metrizable space.

Proof. To show that (1) is equivalent to (2), it suffices to show (2) implies (1). So, suppose that $C_p(X)$ is analytic. Then X has a countable netweight. Since X is a q-space, it follows that X is in fact a separable metrizable space. Then define the metric d on C(X) as in the proof of the sufficiency of Theorem 5.7.4. Since $C_d(X)$, $C_k(X)$ and $C_p(X)$ have the same borel structure and $C_d(X)$ is metrizable, then the analyticity of the Borel structure of $C_p(X)$ implies the analyticity of the topological space $C_d(X)$. Since the identity map id: $C_d(X) \to C_k(X)$ is continuous, then $C_k(X)$ is an analytic topological space. The proof that (3) is equivalent to (1) follows from Theorem 5.7.4 and the q-space hypothesis. ■

<u>Corollary 5.7.6.</u> If X is locally compact, then the following are equivalent.

(1) $C_k(X)$ is Polish.

(2) $C_k(X)$ is analytic.

(3) $C_k(X)$ has countable netweight.

(4) X is Polish.

<u>8. Exercises and Problems for Chapter V.</u>

1. Complete Metrizability. Let X and R be spaces.

(a) $C_k(X,R)$ is completely metrizable if and only if X is a hemicompact k-space and R is completely metrizable.

(b) $C_p(X,R)$ is completely metrizable if and only if X is a countable discrete space and R is completely metrizable.

2. Baire Spaces.

(a) Let $p \in \beta\omega \setminus \omega$, and let $X = \omega \cup \{p\}$. Then $C_p(X)$ is a Baire space.

(b) (Pytkeev [1985], Tkačuk [1986]) If $\{X_t: t \in T\}$ is a family of spaces, then $\Pi\{C_p(X_t): t \in T\}$ is a Baire space if and only if each $C_p(X_t)$ is a Baire space.

(c) (Lutzer and McCoy [1980]) The space $C_p(X)$ is a Baire space if and only if $C_p(Y)$ is a Baire space for every subspace Y of X.

3. Pseudocomplete Spaces. (Lutzer and McCoy [1980], Pytkeev [1985], Tkačuk [1986]) A space is pseudocomplete provided that it has a sequence $\{\mathcal{B}_n: n \in \omega\}$ of π-bases such that if $B_n \in \mathcal{B}_n$ and $\bar{B}_{n+1} \subset B_n$ for each n, then $\cap\{B_n: n \in \omega\}$ is nonempty. The space $C_p(X)$ is pseudocomplete if and only if every pairwise disjoint sequence of finite subsets of X is strongly discrete (see Exercise 2.a). This is equivalent to every countable subset of X being closed and C-embedded in X.

4. Pseudocompact Spaces. (Arhangelskii and Tkačuk [1985]) The space $C_p(X,[0,1])$ is pseudocompact if and only if every countable subset of X is closed and C^*-embedded in X (compare with Exercise 3).

5. Almost Čech-complete and Pseudocomplete σ-space (McCoy and Ntantu 1986). A space is almost Čech-complete provided that it contains a dense Čech-complete subspace. This is stronger than being pseudocomplete (see Exercise 3 for the definition). Recall that a σ-space is a space that has a σ-discrete network. A useful fact is that a Baire σ-space contains a metrizable dense G_δ-subspace. The following are equivalent (cf. Theorem 5.2.1).

(a) $C_\alpha(X)$ is almost Čech-complete.

(b) $C_\alpha(X)$ is a pseudocomplete σ-space.

(c) $C_\alpha(X)$ is completely metrizable.

6. Countably Compact Spaces. (Tani [1979], Arhangelskii [1980]))

(a) If X is a k_R-space or a countably compact space and R is a metric space, then the closure of every countably compact subspace of $C_p(X,R)$ is compact.

(b) The space $C_p(X,[0,1])$ is countably compact if and only if every G_δ-subset of X is open (i.e., X is a P-space).

7. Dense σ-compact Subsets. (Arhangelskii [1980]) The space $C_p(X)$ contains a dense σ-compact subset if and only if $C_p(X)$ has a compact subset that separates points of X.

8. Exact Spaces. (Arhangelskii [1976]) A space X is <u>exact</u> provided that every countably compact subset A of X satisfies:

(i) $\overline{A}$ is compact;

(ii) $\overline{A}$ has countable tightness; and

(iii) every separable subspace of $\overline{A}$ has a countable network.

Note that every metric space is exact.

(a) Every compact exact space is a Fréchet space.

(b) Every subspace of an exact space is exact.

(c) The countable product of exact spaces is exact.

(d) If f: X $\rightarrow$ Y is a continuous bijection and Y is exact, then X is exact.

(e) If X is exact and A $\subset$ X, then the following are equivalent:

(i) A is compact.

(ii) A is sequentially compact.

(iii) A is countably compact.

(f) Let X be almost σ-compact, and let R be metrizable. If τ is any (regular) topology on C(X,R) which is finer than or equal to the topology of pointwise convergence, then $C_\tau(X,R)$ is exact.

9. Borel Subsets (Dijkstra, Grilliot, Lutzer and van Mill [1985], Lutzer, van Mill and Pol [1985]).

(a) If $C_p(X)$ is an F_σ-subset of $\mathbb{R}^X$, then X is discrete.

(b) If $C_p(X)$ is a $G_{\delta\sigma}$-subset of $\mathbb{R}^X$, then X is discrete.

(c) If X is a countable metric space, then $C_p(X)$ is an $F_{\sigma\delta}$-subset of $\mathbb{R}^X$ (and is hence a $G_{\sigma\delta\sigma}$-subset of $\mathbb{R}^X$).

10. Fine Topology (cf. Exercises I.3.1 and II.6.6) Let ρ be a complete metric on R.

(a) If X is a k-space, then $C_{f_\rho}(X)$ is a Baire space.

(b) If X is normal, then the following are equivalent.

(i) $C_{f_\rho}(X)$ is first countable.

(ii) $C_{f_\rho}(X)$ is Lindelöf.

(iii) $C_{f_\rho}(X)$ is separable and completely metrizable.

(iv) X is compact and metrizable.

(d) If X is a nowhere locally compact metric space, then $C_{f_\rho}(X)$ is totally disconnected.

11. Homeomorphic Function Spaces.

(a) (Arhangelskii [1982]) If $C_p(X)$ is linearly homeomorphic to $C_p(Y)$, then $C_k(X)$ is homeomorphic to $C_k(Y)$.

(b) It is a theorem from infinite-dimensional topology (see Bessaga and Pelczynski [1975], Toruńczyk [1981]) that every two infinite-dimensional, locally convex, completely metrizable, topological vector spaces which have the same density are homeomorphic. Deduce that if X and Y are infinite hemicompact k-spaces, then $C_k(X)$ is homeomorphic to $C_k(Y)$ if and only if $w(X) = w(Y)$. It follows that if X is an infinite hemicompact submetrizable k-space, then $C_k(X)$ is homeomorphic to $\mathbb{R}^\omega$.

12. Topological Games. For each of the following, the range space in C(X) is $\mathbb{R}$.

(a) (Gerlits and Nagy [1982], McCoy and Ntantu [1986a]) The Gruenhage game $\Gamma_G(X,x)$ is an infinite two person game played on a space X at a point $x \in X$. On the nth play, player I chooses an open neighborhood U_n of x and then player II chooses a point x_n in U_n. Player I wins if the sequence (x_n) converges to x in X, and otherwise player II wins. If X is homogeneous, then the choice of point x is immaterial; so the game is denoted by $\Gamma_G(X)$ in this case. The space X is called a W-space provided that for each $x \in X$, player I has a winning strategy in $\Gamma_G(X,x)$. (Informally, a strategy for player I is a function from the set of partial plays of the game to the set of open neighborhoods of x. This predetermines the choice for player I in every possible situation.) Every first countable space is a W-space, and every W-space is a Fréchet space. There is a dual game $\Gamma_k(X)$ played on X as follows. On the nth play, player I chooses a compact subset A_n of X and then player II chooses an open set U_n which contains A_n. Player I wins if $\{U_n: n \in \omega\}$ is a k-cover of X, and otherwise player II wins. Player I has a winning strategy in $\Gamma_G(C_k(X))$ if and only if player I has a winning strategy in $\Gamma_k(X)$. If compact sets are replaced by points in the definition of game $\Gamma_k(X)$ (and player I wins if $\{U_n: n \in \omega\}$ is an ω-cover), then player I has a winning strategy in this game on X if and only if player I has a winning strategy in $\Gamma_G(C_p(X))$.

(b) (Lutzer and McCoy [1980], Pytkeev [1985]). The Banach-Mazer game $\Gamma_{BM}(X)$ is played on X with the following rules. Players I and II take turns choosing open sets U_n and V_n, respectively, such that each $V_n \subset U_n$ and each $U_{n+1} \subset V_n$. Then player I wins if $\cap\{U_n: n \in \omega\}$ is empty, and otherwise player II wins. The space X is a Baire space if and only if player I does not have a winning strategy. If player II has a winning strategy then X is called a weakly α-favorable space. The dual game $\Gamma_p(X)$ is played as follows. Players I and II take turns choosing finite sets S_{2n-1} and S_{2n}, respectively, in such a way that S_i and S_j are disjoint for $i \neq j$. Player I wins if $\cup\{S_{2n-1}: n \in \omega\}$ is not strongly discrete (i.e., A is strongly discrete if each $a \in A$ has a neighborhood U_a such that $\{U_a: a \in A\}$ is discrete). Player I has a winning strategy in $\Gamma_{BM}(C_p(X))$ if and only if player I has a winning strategy in $\Gamma_p(X)$. In addition, player II has a winning strategy in $\Gamma_{BM}(C_p(X))$ if and only if every countable subset of

X is closed and C-embedded in X.

13. Function Spaces on the Cantor Set. The properties of a function space can be related to the properties of the range when the domain is fixed. The most natural thing to fix the domain to be (for the compact-open topology) is the Cantor set, K. Let R be an arbitrary space.

(a) $C_k(K,R)$ is (completely) metrizable if and only if R is (completely) metrizable.

(b) $C_k(K,R)$ is 0-dimensional if and only if R is 0-dimensional.

(c) $w(C_k(K,R)) = w(R)$.

(d) $d(C_k(K,R)) = d(R)$.

(e) $\chi(C_k(K,R)) = \sup\{\chi(R,A) : A$ is a compact metrizable subset of $R\}$.

(f) (Michael [1966]) $nw(C_k(K,R)) = knw(C_k(K,R)) = nw(k) = knw(X)$, where k is the space of compact subsets of X with the Vietoris topology.

(g) If R is of first category in itself, then $C_k(K,R)$ is of first category in itself.

(h) If R has no isolated points then $C_k(K,R)$ contains no compact neighborhood.

(i) If R is the space of irrationals, then $C_k(K,R)$ is homeomorphic to R.

14. Set-open Topologies on R^X. Let $R = \mathbb{R}$ and let α be a hereditarily closed, compact network on X. For each $A \in \alpha$ and each bounded open interval V in R, define $[A,V]_F = \{f \in R^X : \overline{f(A)} \subset\}$. The collection $\{[A,V]_F : A \in \alpha$ and V is a bounded open interval in R$\}$ is a subbase for a topology on R^X, which is denoted by R

(a) The space R^X_α is a pseudocomplete Tychonoff space.

(b) The space $C_\alpha(X)$ is a dense subspace of R^X_α.

(c) If $C_\alpha(X)$ contains a dense G_δ-subset of R^X_α, then X is an α_R-space which can be written as a topological sum of a discrete space and a space which is a countable union of members of α.

HISTORICAL NOTES

<u>Chapter 1.</u> The idea of topologizing the set of continuous functions from one topological space into another topological space arose from the notions of pointwise and uniform convergence of sequences of functions. Apparently the work of Ascoli [1883], Arzelà [1889] and Hadamard [1898] marked the beginning of function space theory. The topology of pointwise convergence and the topology of uniform convergence are among the first function space topologies considered in the early years of general topology. The supremum metric topology was studied in Fréchet [1906]. The paper of Tychonoff [1935] showed that the (Tychonoff) product on the set R^X is nothing but the topology of pointwise convergence. In 1945, Fox [1945] defined the compact-open topology. Shortly thereafter, Arens [1946] studied this topology, which he called k-topology. Among other things which Arens proved was the compact-open topology version of Theorem 1.2.3. Set-open topologies in a more general setting were studied by Arens and Dugundji [1951] in connection with the concepts of admissible and proper topologies. Theorem 1.2.5 is due to Jackson [1952], and Example 1.2.7 can be found in Dugundji [1968].

<u>Chapter 2.</u> Admissible (i.e., conjoining) topologies were introduced by Arens [1946] and splitting (i.e., proper) topologies were studied by Arens and Dugundji [1951], where they proved Theorem 2.5.3. Proofs of Theorem 2.5.2 and Corollary 2.5.4.a can be found in Fox [1945]. Corollary 2.5.7 is apparently due to Jackson [1952]; and Morita [1956] proved Corollary 2.5.8. The exponential map was also studied by Brown [1964]. The Whitehead Theorem (Theorem 2.5.10) is in Whitehead's paper [1948]. Most of the results in this chapter are natural and straightforward to prove, and many have occurred in one form or another in different settings. Engelking [1977] includes in his book many of the theorems on diagonal functions, composition functions, product functions and sum functions.

Chapter 3. The notion of continuous convergence was introduced by Arens and Dugundji [1951], and Theorems 3.1.2 and 3.1.3 are due to them. The version of the Ascoli Theorem given in Theorem 3.2.6 can be found in Kelley [1955]. The concepts of hyper- and hypo-Ascoli topologies are discussed in Noble [1969]. Other papers with versions of the Ascoli Theorem in them include: Meyers [1946], Gale [1950], Weston [1959], Poppe [1965], Bagley and Yang [1966], Kaul [1969], Fox and Morales [1973], Yang [1973], Henry, Reynolds and Trapp [1982], and Papadopoulos [1986].

Chapter 4. Cardinal functions in general are studied in the books by Juhász [1971] and [1980]. For a more recent survey on cardinal functions see Hodel [1984] and Juhász [1984]. Since cardinal functions are in some sense generalizations of topological properties, then many of the results in this chapter should be seen as extensions of known theorems, which are stated as corollaries. For instance, Corollary 4.1.3 is a major theorem in Michael [1966]. The first characterization of the density character of $C_\alpha(X)$ for a compact network α is due to Noble [1974]. His result generalizes the countable version (i.e., separability) found in Warner [1958]. The separability of $C_\alpha(X)$ was also studied by Vidossich [1969] and [1970]. Theorem 4.2.4 can be found in Comfort's paper [1971], for example. The concept of total m-boundedness is defined in Comfort and Grant [1981] and Arhangelskii [1981] who characterized total m-bounded groups as being subgroups of groups with cellularity not exceeding m. Corollary 4.2.7 is proved in Ntantu [1985]. The pseudocharacter of function spaces was investigated by Guthrie [1974]. The equivalence of (b), (c) and (d) in Theorem 4.4.2, for the compact-open topology, first occurred in Arens [1946]. For the topology of pointwise convergence, this was done in Fort [1951]. Corollary 4.7.2 can be found in McCoy [1980b], Gerlits and Nagy [1982], and Arhangelskii [1982]; the former two also contain versions of Theorem 4.7.4 for the topology of pointwise convergence. Partial versions of Theorems 4.8.1 and 4.8.3 can be found in Zenor [1980]. A complete version of Theorem 4.8.3 is in Okuyama [1981]. Other inequalities between cardinal functions of function spaces can also be found in Okuyama [1981].

Chapter 5. Uniform completeness has been used in functional analysis for some time, and Theorem 5.1.1 is well-known; the proof can be found for example in Warner [1958]. The equivalence of (b) and (c) in Corollary 5.2.2 can be found for example in Beckenstein, Narici and Suffel [1977]. Corollary 5.2.3 is proved in Lutzer and McCoy [1980]. Also this paper contains a partial characterization of $C_p(X)$ being a Baire space. The full characterization, Theorem 5.3.8, was later given simultaneously by van Douwen [1985] and Pytkeev [1985]. A version of Theorem 5.4.2 can be found for example in Klee [1955], where X is separable and A is compact; but the method of proof is entirely different. A more specialized version of this theorem, which does use the completeness of a function space, occurs in Fox [1941]. The Grothendieck Theorem, 5.5.6, can be found in Grothendieck [1952]. This was generalized in Pryce [1971], and put into a proper topological setting by Arhangelskii [1976]. The Stone-Weierstrass Theorem, 5.6.1, has a long history and plays a role in many aspects of analysis. Theorem 5.6.2 is an example of how this theorem can be used to study function spaces. Theorem 5.6.5 appears in Lutzer and McCoy [1980] and in Lutzer, van Mill and Pol [1985]; and Theorem 5.6.6 is in the former of these papers. The main ideas of section 7 are contained in Christensen [1974]. Theorem 5.7.4 appears in Calbrix [1985]. Theorem 5.7.5 and Corollary 5.7.6 are in Ntantu [1985].

BIBLIOGRAPHY

Atlas O.T.

[1980] "Normal and function spaces", Top. vol I, Colloq. Math. Soc. Janos Bolyai, 23, North-Holland, Amsterdam, 29-33.

Alster K. and Pol R.

[1980] "On function spaces of compact subspaces of Σ-products of the real line", Fund. Math 107, 135-143.

Arens R.

[1946] "A topology of spaces of transformations", Annals of Math. 47, 480-495.

[1952] "Extensions of functions on fully normal spaces", Pacific. J. Math. 2, 11-22.

Arens R. and Dugundji J.

[1951] "Topologies for function spaces", Pacific J. Math. 1, 5-31.

Arhangelskii A.V.

[1966] "Mappings and spaces", Russian Math. Surveys 21:4, 115-162.

[1976] "On some topological spaces that occur in functional analysis", Russian Math. Surveys 31:5, 14-30.

[1978] "On spaces of continuous functions in the topology of pointwise convergence", Soviet Math. Dokl. 19:3, 605-609.

[1980] "Relations among the invariants of topological groups and their subspaces", Russian Math. Surveys 35:3, 1-23.

[1981] "Classes of topological groups", Russian Math. Surveys 36:3, 151-174.

[1982] "On relationships between topological properties of X and $C_p(X)$", Gen. Top. and Appl. to Mod. Anal. and Alg., Proc. of 5th Prague Top. Symp., 24-36.

[1982a] "Factorization theorems and functions spaces: stability and monolithicity", Soviet Math. Dokl. 26, 177-181.

[1983a] "Functional tightness, Q-spaces and τ-embeddings", Comment. Math. Univ. Carolinae 24:1, 105-120.

[1983b] "Function spaces and conditions of completeness type", Vestnik Mosk. Univ. Math. 38:6, 4-9.

[1983c] "Topological properties of function spaces: duality theorems", Soviet Math. Dokl. 27:2, 470-473.

[1984] "Continuous mappings, factorization theorems , and function spaces", Trans. Moscow Math. Soc. 47, 1-22.

[1986] "Hurewicz spaces, analytic sets, and fan tightness of function spaces", Soviet Math. Dokl. 33, 396-399.

Arhangelskii A.V. and Tkačuk V.V.

[1985] "Function spaces and topological invariants", (preprint).

Arzela C.

[1889] "Funzioni di linee", Atti della Reale Accademia dei Lincei, Rendiconti 5, 342–348.

Ascoli G.

[1883] "Le curve limite di una varieta data di curve", Mem. Accad. Lincei (3) 18, 521–586.

Bagley R.W. and Yang J.S.

[1966] "On k-spaces and function spaces", Proc. Amer. Math. Soc. 17, 703–705.

Balogh Z.

[1984] "On hereditarily strong Σ-spaces", Top. and Appl. 17, 199–215.

Beckenstein E., Narici L. and Suffel C.

[1977] "Topological algebras", Notas de Mat. 60, North-Holland, N.Y.

Beer G.

[1983] "On uniform convergence of continuous functions and topological convergence of sets", Can. Math. Bull. 26, 418–424.

[1985] "More on convergence of continuous functions and topological convergence of sets", Can. Math. Bull. 28, 52–59.

[1986] "On a generic optimization theorem of Petar Kenderov", (preprint).

Bessaga C. and Pelczynski A.

[1975] "Infinite-dimensional topology", P.W.N., Warszawa.

Borges C.

[1966] "On stratifiable spaces", Pacific J. Math. 11, 1–16.

[1966a] "On function spaces of stratifiable spaces and compact spaces", Proc. Amer. Math. Soc. 17, 1074–1078.

[1979] "Compact-open verses k-compact-open", Proc. Amer. Math. Soc. 73, 129–133.

Brown R.

[1964] "Function spaces and product topologies", Quart. J. Math. Oxford (2) 15, 238–250.

Calbrix J.

[1985] "Espaces K_σ et espaces des applications continues", Bull. Soc. Math. France 113, 183–203.

Christensen J-P.R.

[1974] "Topology and Borel structure", North-Holland, Amsterdam.

Comfort W.W.

[1971] "A survey of cardinal invariants", Top. and Appl. 1, 163-199.

Comfort W.W. and Grant D.L.

[1981] "Cardinal invariants, pseudocompactness and minimality: some recent advances in the topological theory of topological groups", Top. Proc. 6, 227-265.

Comfort W.W. and Hager A.W.

[1970] "Estimates for the number of real-valued continuous functions", Trans. Amer. Math. Soc. 150, 619-631.

Corson H.H.

[1959] "Normality in subsets of product spaces", Amer. J. Math. 81, 785-796.

Corson H.H. and Lindenstrauss

[1966] "On function spaces which are Lindelöf spaces", Trans. Amer. Math. Soc. 121, 476-491.

Dijkstra J., Grilliot T., Lutzer D. and Van Mill J.

[1985] "Function spaces of low Borel complexity", Proc. Amer. Math. Soc. 94, 703-710.

van Douwen E.K.

[1985] Private communication.

Dugundji J.

[1951] "An extension of Tietze's theorem", Pacific J. Math. 1, 353-367.

[1968] "Topology", Allyn and Bacon, Inc., Boston.

Eklund A.D.

[1978] "The fine topology and other topologies on C(X,Y)", Dissertation, Virginia Polytechnic Institute and State University, Blacksburg, Virginia.

Engelking R.

[1977] "General Topology", P.W.N., Warszawa.

Fadell E.

[1959] "B paracompact does not imply B^I paracompact", Proc. Amer. Math. Soc. 9, 839-840.

Fort M.K.

[1951] "A note on pointwise convergence", Proc. Amer. Math. Soc. 2, 34-35.

Fox R.H.

[1945] "On topologies for function spaces", Bull. Amer. Math. Soc. 51, 429–432.

Fox G. and Morales P.

[1973] "A non-Hausdorff Ascoli theorem for k_3-spaces", Proc. Amer. Math. Soc. 39, 633–636.

Fréchet M.

[1906] "Sur quelques points du calcul functionnel", Rend. del Circ. Mat. di Palermo, 1–74.

Fuller R.V.

[1972] "Condition for a function space to be locally compact", Proc. Amer. Math. Soc. 36, 615–617.

Gale D.

[1950] "Compact sets of functions and function rings", Proc. Amer. Math. Soc. 1, 303–308.

Gerlits J.

[1983] "Some properties of C(X), II", Top. and Appl. 15, 255–262.

Gerlits J. and Nagy Zs.

[1982] "Some properties of C(X), I", Top. and Appl. 14, 151–161.

Gillman L. and Jerison M.

[1960] "Rings of continuous functions", Van Nostrand, Princeton N.J.

Grothendieck A.

[1952] "Critères de compacité dans les espaces fonctionnels généraux", Amer. J. Math. 74, 168–186.

Gul'ko S.P.

[1977] "On properties of subsets of Σ-products", Soviet Math. Dokl. 18, 1438–1442.

[1978] "On the properties of some function spaces", Soviet Math. Dokl., 1420–1424.

[1979] "On the structure of spaces of continuous functions and their hereditary paracompactness", Uspekhi Mat. Nauk 34:6, 33–40.

Guthrie J.A.

[1971] "A characterization of $\aleph_0$-spaces", Gen. Top. and Appl. 1, 105–110.

[1973] "Mapping spaces and cs-networks", Pacific J. Math. 47, 465–471.

[1974] "Ascoli theorems and the pseudocharacter of mapping spaces", Bull. Austral. Math. Soc. 10, 403–408.

Hadamard J.

[1898] "Sur certaines applications possibles de la théorie des ensembles", Verhandl. Ersten Intern. Math. Kongresses, B.G. Teubner, Leipzig.

Hager A.W.

[1969] "Approximation of real continuous functions on Lindelöf spaces", Proc. Amer. Math. Soc. 22, 156–163.

Hansard J.D.

[1970] "Function space topologies", Pacific J. Math. 35, 381–388.

Heath R.W., Lutzer D.J. and Zenor P.L.

[1975] "On continuous extenders", Studies in Topology, Academic Press, N.Y., 203–213.

Helmer D.

[1981] "Criteria for Eberlein compactness in spaces of continuous functions", Manuscripta Math. 35, 27–51.

Henry M., Reynolds D. and Trapp G.

[1982] "A note on Gale's property G", Top. Proc. 7, 193–196.

[1985] "Equicontinuous and regular collections of functions", Top. Proc.

Hodel R.

[1984] "Cardinal functions I", Handbook of Set-theoretic Topology, North-Holland, Amsterdam, 1–61.

Isbell J.R.

[1964] "Uniform spaces", Math. Surveys no. 12, Amer. Math. Soc., Providence, R.I.

Irudayanathan A. and Naimpally S.

[1966] "connected open topology for function spaces", Indag. Math. 28, 22–24.

Jackson J.R.

[1952] "Spaces of mappings on topological products with appliances to homotopy theory", Proc. Amer. Math. Soc. 3, 327–333.

Jeschek F.

[1971] "Remarks on 'connected' topologies for functions spaces", Bull. Acad. Polon. Sci., Ser. Sci. Math. Astron. Phys. 29, 1045–1051.

Juhász I.

[1971] "Cardinal functions in topology", Mathematisch Centrum, Amsterdam.

[1980] "Cardinal functions in topology – ten years later", Mathematisch Centrum Tracts 123, Amsterdam.

[1984] "Cardinal functions II", Handbook of Set-theoretic Topology, North-Holland, Amsterdam, 63–109.

Kaul S.K.

[1969] "Compact subsets in function spaces", Bull. Canad. Math. 12, 461–466.

Kelley J.L.

[1955] "General Topology", Van Nostrand, New York.

Krikorian, N.

[1969] "A note concerning the fine topology on function spaces", Compositio Math. 21, 343–348.

Krivorucko A.I.

[1972] "On the cardinality of the set of continuous functions", Soviet Math. Dokl. 13, 1364–1367.

[1973] "On cardinal invariants of spaces and mappings", Soviet Math. Dokl. 14, 1642–1647.

[1975] "The cardinality and density of spaces of mappings", Soviet Math. Dokl. 16, 281–285.

Lambrinos, P

[1980] "Boundedly generated topological spaces", Manuscripta Math. 31, 425–438.

[1981] "The bounded-open topology on function spaces", Manuscripta Math. 36, 47–66.

Lehner W.

[1978] "Über die Bedeutung gewisser Varianten des Baire'schen Kategorien-begriffs für die Funktionenräume $C_c(T)$", Dissertation, Ludwig-Maximilian-Universität, München.

Lutzer D.J. and McCoy R.A.

[1980] "Category in function spaces I", Pacific J. Math. 90, 145–168.

Lutzer D., van Mill J. and Pol R.

[1985] "Filters and the descriptive complexity of function spaces", (preprint).

McCoy R.A.

[1978] "Characterization of pseudocompactness by the topology of uniform convergence on function spaces", J. Austral. Math. Soc. 26, 251–256.

[1978a] "Submetrizable spaces and almost σ-compact function spaces", Proc. Amer. Math. Soc. 71, 138–142.

[1980] "Countability properties of function spaces", Rocky Mountain J. Math 10, 717–730.

[1980a] "Necessary conditions for function spaces to be Lindelöf", Glasnik Mat. 15, 163–168.

[1980b] "k–space function spaces", Intern. J. Math and Math. Sci. 3, 701–711.

[1980c] "Function spaces which are k–spaces", Top. Proc. 5, 139–146.

[1980d] "The evaluation identification in function spaces", Top. and Appl. 11, 189–197.

[1986] "The evaluation identification and the Ascoli theorem", Indian J. Math. 28.

[1986] "Fine topology on function spaces", Intern. J. Math. and Math. Sci. 9, 417–424.

McCoy R.A. and Ntantu I.

[1986] "Completeness properties of function spaces", Top. and Appl. 22, 191–206.

[1986a] "Countability properties of functin spaces with set–open topologies", Top. Proc. 10.

Meyer P.R.

[1964] "Topologies of spaces of real–valued functions", Dissertation, Columbia University, New York.

[1967] "Topologies with the Stone–Weierstrass property", Trans. Amer. Math. Soc. 126, 236–243.

[1970] "Function spaces and the Aleksandrov–Urysohn conjecture", Estratto dagli Annali di Matematia Pura ed Applicata, Ser. 4, 86, 25–29.

Meyers S.B.

[1946] "Equicontinuous sets of mappings", Ann. Math. 47, 496–502.

[1949] "Spaces of continuous functions", Bull. Amer. Math. Soc. 55, 402–407.

Michael E.

[1951] "Topologies on spaces of subsets", Trans. Amer. Math. Soc. 71, 152–182.

[1953] "Some extension theorems for continuous functions", Pacific J. Math. 3, 789–806.

[1956] "On a theorem of Rudin and Klee", Proc. Amer. Math. Soc. 12, 921.

[1966] "$\aleph_0$–spaces", J. Math. Mech. 15, 983–1002.

[1977] "$\aleph_0$–spaces and a function space theorem of R. Pol", Indiana Univ. Math J. 26, 299–306.

Morita K.

[1956] "Note on mapping spaces", Proc. Japan Acad. 32, 671–675.

Morris P.D.

[1966] "Spaces of continuous functions on dispersed sets", Dissertation, University of Texas, Austin, Texas.

Nachbin L.

[1954] "Topological vector spaces of continuous functions", Proc. Nat. Acad. Sci. USA 40, 471–474.

Naimpally S.A.

[1966] "Graph topology for function spaces", Trans. Amer. Math. Soc. 123, 267–272.

Naimpally S.A. and Pareek C.M.

[1970] "Graph topologies for function spaces, II", Annales Soc. Math. Pol. Series I, 13, 222–231.

Namioka I.

[1974] "Separate continuity and joint continuity", Pacific J. Math. 51, 515–531.

Noble N.

[1969] "Ascoli theorems and the exponential map", Trans. Amer. Math. Soc. 143, 391–411.

[1969a] "Products with closed projections", Trans. Amer. Math. Soc. 140, 381–391.

[1974] "The density character of function spaces", Proc. Amer. Math. Soc. 42, 228–233.

Ntantu I

[1985] "The compact-open topology on C(X)", Dissertation, Virginia Polytechnic Institute and State University, Blacksburg, Virginia.

[1986a] "On Cardinal functions related to function spaces", (preprint).

[1986b] "Cardinal functions on hyperspaces and function spaces", (preprint).

Okuyama A.

[1981] "Some relationships between function spaces and hyperspaces by compact sets", Gen. Top. and Rel. to Mod. Anal. and Alg. V, Proc. Fifth Prague Top. Symp. 527–535.

[1986] "On a topology of the set of linear continuous functionals", Kobe J. Math. 3, 213–217.

O'Meara P.

[1971] "Paracompactness in function spaces with the compact-open topology", Proc. Amer. Math. Soc. 29, 183–189.

Pavlovskii D.S.

[1979] "Spaces of open sets and spaces of continuous functions", Soviet Math. Dokl. 20, 564–568.

Pelczynski A. and Semadeni Z.

[1959] "Spaces of continuous functions (III) (spaces C(Ω) for Ω without perfect subsets)", Studia Math. 18, 211–222.

Pervin W.J.

[1967] "On the connected-open topology", Indag. Math. 29, 126–127.

Pol. R.

[1974] "Normality in function spaces", Fund. Math. 84, 145–155.

[1979] "A function space C(X) which is weakly Lindelöf but not weakly compactly generated", Studia Math. 69, 279–285.

[1980] "A theorem on the weak topology of C(X) for compact scattered X", Fund. Math. 106, 135–140.

Poppe H.

[1965] "Stetige Konvergenz und der Satz von Ascoli und Arzela", Math. Nachr. 30, 87–122.

[1966] "Ein Kompaktheitskriterium für Abbildungsräume mit einer verallgemeinerten uniformen Struktur", Gen. Top. and Rel. to Mod. Anal. and Alg. II, Proc. Second Prague Top. Symp., 284–289.

[1967] "Über Graphentopologien für Abbildungsräume I", Bull. Acad. Pol. Sci., Ser. Sci. Math. Astron. Phys. 15, 71–80.

[1968] "Über Graphentopologien für Abbildungsräume II", Math. Nachr. 38, 89–96.

[1970] "Compactness in function spaces with a generalized uniform structure II", Bull. Acad. Pol. Sci., Ser. Sci. Math. Astron. Phys. 18, 567–573.

Pryce J.D.

[1971] "A device of R.J. Whitley's applied to pointwise compactness in spaces of continuous functions", Proc. London Math. Soc. 23, 532–546.

Pytkeev E.G.

[1982] "On sequentiality of spaces of continuous functions", Communications Moscow Math. Soc. 190–191.

[1985] "The Baire property of spaces of continuous functions", Math. Zametki 38, 726–740.

Rajagopalan M. and Wheeler R.F.

[1976] "Sequential compactness of X implies a completeness property for C(X)", Canad. J. Math. 28, 207–210.

Rudin M.E. and Klee V.L.

[1956] "A note on certain functin spaces", Arch. Math. 7, 469–470.

Sakai M.

[1988] "On property C′′ and function spaces", to appear.

Shirota T.

[1954] "On locally convex vector spaces of continuous functions", Proc. Japan Acad. 30, 294–298.

Siska J.

[1982] "The LCC-topology on the space of continuous functions", Comment. Math. Univ. Carolinae 23, 89–103.

Sokolov G.A.

[1984] "On some classes of compact spaces lying in Σ-products", Comment. Math. Univ. Carolinae 25, 219–231.

Stone M.H.

[1947] "The generalized Weierstrass approximation theorem", Math. Mag. 21, 167–183 and 237–254.

Talagrand M.

[1977] "Sur les espaces de Banach faiblement k-analytiques", Comptes Rendu Acad. Sci. Serie A 285, 119–122.

Tani T.

[1979] "Perfectly finally compact spaces are hard", Math. Japonica 24, 323–326.

Tkačuk V.V.

[1984] "On the multiplicity of certain properties of spaces of mappings in the topology of pointwise convergence", Vest. Mosk. Univ. Mat. 39, 53–57.

[1986] "The spaces $C_p(X)$: decomposition into a countable union of bounded subspaces and completeness properties", (preprint).

Toruńczyk H.

[1981] "Characterizing Hilbert space topology", Fund. Math. 111, 247–262.

Tychonoff A.

[1935] "Über einer Funktionenräum", Math. Ann. 111, 762–766.

Uspenskii V.V.

[1978] "On embeddings in function spaces", Soviet Math. Dokl. 19, 1159–1162.

[1982] "On the frequency spectrum of function spaces", Vestnik Moskov. Univ. Ser. I Mat. Mekh., 31–35.

[1983] "A characterization of real compactness in terms of the topology of pointwise convergence on the function space", Comment. Math. Univ. Carolinae 24, 121–126.

Velichko N.V.

[1981] "Weak topology of spaces of continuous functions", Mat. Zametki 30, 703–712.

[1982] "On the theory of spaces of continuous functions", Commun. Moscow Math. Soc., 149–150.

Vidossich G.

[1969] "A remark on the density character of function spaces", Proc. Amer. Math. Soc. 22, 618–619.

[1969a] "On topological spaces whose function space is of second category", Invent. Math. 8, 111–113.

[1970] "Characterizing separability of function spaces", Inventiones Math. 10, 205–208.

[1971] "On a theorem of Corson and Lindenstrauss on Lindelöf function spaces", Israel J. Math. 9, 271–278.

[1972] "Function spaces which are pseudo-ℵ-compact spaces", (preprint).

[1972a] "On compactness in function spaces", Proc. Amer. Math. Soc. 33, 594–598.

Warner S.

[1958] "The topology of compact convergence on continuous function spaces", Duke Math. J. 25, 265–282.

Weston J.D.

[1959] "A generalization of Ascoli's theorem", Mathematika 6, 19–24.

Whitehead J.H.C.

[1948] "A note on a theorem due to Borsuk", Bull. Amer. Math. Soc. 54, 1125–1132.

Willard S.

[1970] "General topology", Addision–Wesley Publishing Co.

Yang J.S.

[1973] "Property (G), regularity, and semi–equicontinuity", Canad. Math. Bull. 16, 587–594.

Yoshioka I.

[1980] "Note on topologies for function spaces", Math. Japonica 25, 373–377.

Young N.J.

[1973] "Compactness in functin spaces; another proof of a theorem of J. D. Pryce", J. London Math. Soc. 6, 739–740.

Zenor P.

[1980] "Hereditary m–separability and the hereditary m–Lindelöf property in product spaces and function spaces", Fund. Math. 106, 175–180.

LIST OF SYMBOLS

SUBJECT INDEX

Virginia Polytechnic Institute and State University
Blacksburg, Virginia 24061
Middle Tennessee State University
Murfreesboro, Tennessee 37132

The second author was partially supported by a faculty research grant from Middle Tennessee State University.

Vol. 1145: G. Winkler, Choquet Order and Simplices. VI, 143 pages. 1985.

Vol. 1146: Séminaire d'Algèbre Paul Dubreil et Marie-Paule Malliavin. Proceedings, 1983–1984. Edité par M.-P. Malliavin. IV, 420 pages. 1985.

Vol. 1147: M. Wschebor, Surfaces Aléatoires. VII, 111 pages. 1985.

Vol. 1148: Mark A. Kon, Probability Distributions in Quantum Statistical Mechanics. V, 121 pages. 1985.

Vol. 1149: Universal Algebra and Lattice Theory. Proceedings, 1984. Edited by S. D. Comer. VI, 282 pages. 1985.

Vol. 1150: B. Kawohl, Rearrangements and Convexity of Level Sets in PDE. V, 136 pages. 1985.

Vol 1151: Ordinary and Partial Differential Equations. Proceedings, 1984. Edited by B.D. Sleeman and R.J. Jarvis. XIV, 357 pages. 1985.

Vol. 1152: H. Widom, Asymptotic Expansions for Pseudodifferential Operators on Bounded Domains. V, 150 pages. 1985.

Vol. 1153: Probability in Banach Spaces V. Proceedings, 1984. Edited by A. Beck, R. Dudley, M. Hahn, J. Kuelbs and M. Marcus. VI, 457 pages. 1985.

Vol. 1154: D.S. Naidu, A.K. Rao, Singular Pertubation Analysis of Discrete Control Systems. IX, 195 pages. 1985.

Vol. 1155: Stability Problems for Stochastic Models. Proceedings, 1984. Edited by V.V. Kalashnikov and V.M. Zolotarev. VI, 447 pages. 1985.

Vol. 1156: Global Differential Geometry and Global Analysis 1984. Proceedings, 1984. Edited by D. Ferus, R.B. Gardner, S. Helgason and U. Simon. V, 339 pages. 1985.

Vol. 1157: H. Levine, Classifying Immersions into $\mathbb{R}^4$ over Stable Maps of 3-Manifolds into $\mathbb{R}^2$. V, 163 pages. 1985.

Vol. 1158: Stochastic Processes – Mathematics and Physics. Proceedings, 1984. Edited by S. Albeverio, Ph. Blanchard and L. Streit. VI, 230 pages. 1986.

Vol. 1159: Schrödinger Operators, Como 1984. Seminar. Edited by S. Graffi. VIII, 272 pages. 1986.

Vol. 1160: J.-C. van der Meer, The Hamiltonian Hopf Bifurcation. VI, 115 pages. 1985.

Vol. 1161: Harmonic Mappings and Minimal Immersions, Montecatini 1984. Seminar. Edited by E. Giusti. VII, 285 pages. 1985.

Vol. 1162: S.J.L. van Eijndhoven, J. de Graaf, Trajectory Spaces, Generalized Functions and Unbounded Operators. IV, 272 pages. 1985.

Vol. 1163: Iteration Theory and its Functional Equations. Proceedings, 1984. Edited by R. Liedl, L. Reich and Gy. Targonski. VIII, 231 pages. 1985.

Vol. 1164: M. Meschiari, J.H. Rawnsley, S. Salamon, Geometry Seminar "Luigi Bianchi" II – 1984. Edited by E. Vesentini. VI, 224 pages. 1985.

Vol. 1165: Seminar on Deformations. Proceedings, 1982/84. Edited by J. Ławrynowicz. IX, 331 pages. 1985.

Vol. 1166: Banach Spaces. Proceedings, 1984. Edited by N. Kalton and E. Saab. VI, 199 pages. 1985.

Vol. 1167: Geometry and Topology. Proceedings, 1983–84. Edited by J. Alexander and J. Harer. VI, 292 pages. 1985.

Vol. 1168: S.S. Agaian, Hadamard Matrices and their Applications. III, 227 pages. 1985.

Vol. 1169: W.A. Light, E.W. Cheney, Approximation Theory in Tensor Product Spaces. VII, 157 pages. 1985.

Vol. 1170: B.S. Thomson, Real Functions. VII, 229 pages. 1985.

Vol. 1171: Polynômes Orthogonaux et Applications. Proceedings, 1984. Edité par C. Brezinski, A. Draux, A.P. Magnus, P. Maroni et A. Ronveaux. XXXVII, 584 pages. 1985.

Vol. 1172: Algebraic Topology, Göttingen 1984. Proceedings. Edited by L. Smith. VI, 209 pages. 1985.

Vol. 1173: H. Delfs, M. Knebusch, Locally Semialgebraic Spaces. XVI, 329 pages. 1985.

Vol. 1174: Categories in Continuum Physics, Buffalo 1982. Seminar. Edited by F.W. Lawvere and S.H. Schanuel. V, 126 pages. 1986.

Vol. 1175: K. Mathiak, Valuations of Skew Fields and Projective Hjelmslev Spaces. VII, 116 pages. 1986.

Vol. 1176: R.R. Bruner, J.P. May, J.E. McClure, M. Steinberger, H_∞ Ring Spectra and their Applications. VII, 388 pages. 1986.

Vol. 1177: Representation Theory I. Finite Dimensional Algebras. Proceedings, 1984. Edited by V. Dlab, P. Gabriel and G. Michler. XV, 340 pages. 1986.

Vol. 1178: Representation Theory II. Groups and Orders. Proceedings, 1984. Edited by V. Dlab, P. Gabriel and G. Michler. XV, 370 pages. 1986.

Vol. 1179: Shi J.-Y. The Kazhdan-Lusztig Cells in Certain Affine Weyl Groups. X, 307 pages. 1986.

Vol. 1180: R. Carmona, H. Kesten, J.B. Walsh, École d'Été de Probabilités de Saint-Flour XIV – 1984. Édité par P.L. Hennequin. X, 438 pages. 1986.

Vol. 1181: Buildings and the Geometry of Diagrams, Como 1984. Seminar. Edited by L. Rosati. VII, 277 pages. 1986.

Vol. 1182: S. Shelah, Around Classification Theory of Models. VII, 279 pages. 1986.

Vol. 1183: Algebra, Algebraic Topology and their Interactions. Proceedings, 1983. Edited by J.-E. Roos. XI, 396 pages. 1986.

Vol. 1184: W. Arendt, A. Grabosch, G. Greiner, U. Groh, H.P. Lotz, U. Moustakas, R. Nagel, F. Neubrander, U. Schlotterbeck, One-parameter Semigroups of Positive Operators. Edited by R. Nagel. X, 460 pages. 1986.

Vol. 1185: Group Theory, Beijing 1984. Proceedings. Edited by Tuan H.F. V, 403 pages. 1986.

Vol. 1186: Lyapunov Exponents. Proceedings, 1984. Edited by L. Arnold and V. Wihstutz. VI, 374 pages. 1986.

Vol. 1187: Y. Diers, Categories of Boolean Sheaves of Simple Algebras. VI, 168 pages. 1986.

Vol. 1188: Fonctions de Plusieurs Variables Complexes V. Séminaire, 1979–85. Edité par François Norguet. VI, 306 pages. 1986.

Vol. 1189: J. Lukeš, J. Malý, L. Zajíček, Fine Topology Methods in Real Analysis and Potential Theory. X, 472 pages. 1986.

Vol. 1190: Optimization and Related Fields. Proceedings, 1984. Edited by R. Conti, E. De Giorgi and F. Giannessi. VIII, 419 pages. 1986.

Vol. 1191: A.R. Its, V.Yu. Novokshenov, The Isomonodromic Deformation Method in the Theory of Painlevé Equations. IV, 313 pages. 1986.

Vol. 1192: Equadiff 6. Proceedings, 1985. Edited by J. Vosmansky and M. Zlámal. XXIII, 404 pages. 1986.

Vol. 1193: Geometrical and Statistical Aspects of Probability in Banach Spaces. Proceedings, 1985. Edited by X. Fernique, B. Heinkel, M.B. Marcus and P.A. Meyer. IV, 128 pages. 1986.

Vol. 1194: Complex Analysis and Algebraic Geometry. Proceedings, 1985. Edited by H. Grauert. VI, 235 pages. 1986.

Vol.1195: J.M. Barbosa, A.G. Colares, Minimal Surfaces in $\mathbb{R}^3$. X, 124 pages. 1986.

Vol. 1196: E. Casas-Alvero, S. Xambó-Descamps, The Enumerative Theory of Conics after Halphen. IX, 130 pages. 1986.

Vol. 1197: Ring Theory. Proceedings, 1985. Edited by F.M.J. van Oystaeyen. V, 231 pages. 1986.

Vol. 1198: Séminaire d'Analyse, P. Lelong – P. Dolbeault – H. Skoda. Seminar 1983/84. X, 260 pages. 1986.

Vol. 1199: Analytic Theory of Continued Fractions II. Proceedings, 1985. Edited by W.J. Thron. VI, 299 pages. 1986.

Vol. 1200: V.D. Milman, G. Schechtman, Asymptotic Theory of Finite Dimensional Normed Spaces. With an Appendix by M. Gromov. VIII, 156 pages. 1986.